# *About Island Press*

Island Press, a nonprofit organization, publishes, markets, and distributes the most advanced thinking on the conservation of our natural resources—books about soil, land, water, forests, wildlife, and hazardous and toxic wastes. These books are practical tools used by public officials, business and industry leaders, natural resource managers, and concerned citizens working to solve both local and global resource problems.

Founded in 1978, Island Press reorganized in 1984 to meet the increasing demand for substantive books on all resource-related issues. Island Press publishes and distributes under its own imprint and offers these services to other nonprofit organizations.

Support for Island Press is provided by Apple Computers Inc., Mary Reynolds Babcock Foundation, Geraldine R. Dodge Foundation, The Charles Engelhard Foundation, The Ford Foundation, Glen Eagles Foundation, The George Gund Foundation, William and Flora Hewlett Foundation, The Joyce Foundation, The John D. and Catherine T. MacArthur Foundation, The Andrew W. Mellon Foundation, The Joyce Mertz-Gilmore Foundation, The New-Land Foundation, The J. N. Pew, Jr. Charitable Trust, Alida Rockefeller, The Rockefeller Brothers Fund, The Florence and John Schumann Foundation, The Tides Foundation, and individual donors.

# TREES, WHY DO YOU WAIT?

# TREES, WHY DO YOU WAIT?

*America's Changing Rural Culture*

BY RICHARD CRITCHFIELD

ISLAND PRESS

*Washington, D.C.*  □  *Covelo, California*

Permissions to quote from copyrighted material appear on page 255.

Library of Congress Cataloging-in-Publication Data

Critchfield, Richard.
    Trees, why do you wait? / Richard Critchfield.
        p.   cm.
    Includes index.
        ISBN 1-55963-029-9 (alk. paper). — ISBN 1-55963-028-0 (pbk. : alk. paper)
        1. United States—Rural conditions.   2. Agriculture—Economic aspects—United States.   I. Title.
    HN59.2.C75   1991
    307.72'0973—dc 20                                                   90-20898
                                                                                    CIP

Printed on recycled, acid-free paper

Manufactured in the United States of America

10   9   8   7   6   5   4   3   2

Why do you listen, trees?
Why do you wait?
Why do you fumble at the breeze—
Gesticulate
With hopeless fluttering hands—
Stare down the vanished road beyond the gate
That now no longer stands?
Why do you wait?
Trees—
Why do you listen, trees?

From "The Farm," Archibald MacLeish
*New and Collected Poems, 1917–1982*

# Contents

# *Acknowledgments*

I wish to thank a group of authorities in their fields, whose ideas and comments, many quoted, were of such help to me in this book. Most of them, though known much longer, were interviewed in 1987 and 1990: Norman E. Borlaug, Theodore W. Schultz, William H. McNeill, Gilbert C. Fite, Neil Harl, Vernon Ruttan, Calvin Beale, Wayne Rasmussen, Don Paarlberg, and Lowell S. Hardin.

The second group to whom I am most indebted are the people in the two farm communities themselves. I wish to thank John Morrin, Stanley Goodwin, Charles Eldredge, Mary Eldredge, Chuck Eldredge, Mitchell Lloyd, Clark Lloyd, Albert Seibold, Don Musha, LeVon Kirkeide, Paul Kirkeide, Jim Kirkeide, Ray Klindworth, Jerry Klindworth, John Bollingberg, Kurt Bollingberg, Gerald Hagemeister, Lorraine Hagemeister, Jerome Hagemeister, Jennifer Hagemeister, Mark Hagemeister, David Clough, Cyrus Clough, Terry Olschlager, the late Emma Lorz, William Lorz, Ella Lenz, Norman Weckerly, Betty Melby, Lannis Faleide, Lisa Faleide, the late Mary Carter, Mabel Swanson, Jim Parsons, Rees Price, Alvin Mohr, the late Chester Zumpf, Earl Zumpf, Father John Herron, Rev. Arlyn Anfinrud, Norman Kessler, Dennis Walsh, Sheriff Curtis Pellet, Helen Musha, the late Fred Mietz, Esperanza Tamayo, all in North Dakota. In Iowa: Helen Collins, the late Joe Hayden, the late Beryl Secrist, Paul Secrist, Dorothy Miller, Helen Carlson, Hilda Armstrong, the late Darold Armstrong, Hadwen Collins, Ruth Collins, Johnny Collins, Amy Collins, Steve Preuss, Jim Wild, Lois Wild, Craig Loertscher, Robert Williams, Maxine Williams, Rev. Michael Bahde, Paul Perez, Christine Perez, Dixie Wolfe, Don Wolfe, and in Waterloo, Martha Nash, Ada Tredwell, and Ruth Anderson.

I am also grateful to my two research assistants, Grace Hayek of Iowa City, Iowa, and Chicago, and Tim Holzkamm of Bad Medicine Lake, Minnesota. Ms. Hayek and Mr. Holzkamm had worked with me in 1982–83 on an earlier book; in April–August, 1987, we taped 124 interviews in Iowa, North Dakota, and Minnesota; in 1990 I returned alone and did 52 more, mainly with the same people. Those interviewed and not directly quoted, but who helped with information and analysis, in Iowa were: Robert W. Jolley and Kenneth Stone, Department of Economics, Iowa State University; Lawrence Gelfand and Robert J. Neymeyer, Department of History, University of Iowa; William L. Brown, former president, Pioneer Hi-Bred International, Inc.; Loren Soth, *Des Moines Register;* Bob Davis, Dan Miller, John Peterson, Mike Williams, John Robertson, Myron Williams, *Iowa Today;* Al Schwegel, *Cedar Rapids Gazette;* John Halder, Jon de Neue, Kirkwood Community College, Cedar Rapids; Esther Williams, Vinton; Vicki Collins, James May, Viola; Michael Douglas, George Champlin, Lowell Sovereign, Cresco; George Knaphus, William Murray, Iowa State University, Ames. In North Dakota: Lee Stenjehem, Richard Maine, Wayne Cornelius, Wally Newman, Willis Faul, Karen Stave, Ruth Widiger, Richard Pfeiffer, Fessenden; Keith Peltier, Chris Sellie, Harvey; Dalene Battagler, Bill Battagler, Duane Moen, Betty Moen, Robert Bell, Joel Bell, Kevin Bell, Marguerite Pratt, Carla Pratt, Paul Pratt, Harold Rasmussen, Laura Rasmussen, Hunter; Joe Peltier, Arthur; Douglas Burgum, Fargo; Randal Mikkelsen, *Grand Forks Herald* and *Agweek;* Dean H. Roald Lund, College of Agriculture, North Dakota State University.

Also Janice O'Connell and Jack Holzhueter, State Historical Society of Wisconsin, Madison; Dan and Mary Johnson, Madison; Assistant Secretary of Agriculture for Education and Science Orville Bentley, Washington; Matt McMahon, International Maize and Wheat Improvement Center, Et Batan, Mexico; former Secretary of Agriculture Earl Butz, Purdue University; Dick Youngblood, business editor, *Minneapolis Star and Tribune;* E. C. A. Runge, Texas A&M University; former Secretary of Agriculture Robert Lounsberry, State of Iowa; Lester R. Brown, president, Worldwatch Institute, Washington, D.C. And for their views on religion, Ronald Bowlby, Bishop of Southwark, London, and David Jenkins, Bishop of Durham, Durham, England.

I am also grateful to John Bye of the North Dakota Institute for Regional Studies for again being so helpful, to Norman R. Collins of the Ford Foundation for his support and forbearance, and to Beth A. Beisel, for her skillful production editing. Lastly, I thank Barbara Dean, my editor, with whom it was a great pleasure to work.

RC

The people of Prairie and Crow Creek (*in order of appearance; their names and some place names have been changed*).

## Prairie

## Crow Creek

# I

# *What Is Being Lost?*

A WINDBREAK ON an abandoned homestead where the house and barns have been torn down and just the trees are left is just about the bleakest sight in rural America today. People are giving up their farms. Maybe they simply retire—that is the most common way—or they die or go belly-up or bankrupt. Whatever the reason, homesteads are being bulldozed down and plowed up.

And there's a family that used to live there who used to come to town and buy groceries, fuel, and drugs; visit the doctor; send children to school; marry in the church; and be buried in the graveyard. And that family is not there anymore. There is no family. Old homesteads are being destroyed all over the Midwest. And all that is left is trees.

For rural America, the 1980s, ending in 2 years of drought, were as depressing as the Dust Bowl years of the 1920s. The farm population has been squeezed to below five million. Bewildered, debt ridden, farmers wonder what is going to happen to them.

Fifty, 60 years ago, the rural scene looked much brighter. Before the 1920s, old-timers say, change in American farming was slow. Silent movies of those days wonderfully record dusty dirt roads, farm wagons and Model Ts chugging by, threshers in overalls pitching bundles, small family farms every quarter section or so, with cows, pigs, and chickens. It was a way of life whose speed and power was set by the 3-mile-an-hour gait of the horse.

By 1940, as highly mechanized, highly capitalized farming took over, this way of life was already becoming a nostalgic memory. Since 1940 the number of Americans who farm has dropped from about 30 percent to scarcely 2 percent. In the same 50 years, the country saw the end of the

3

Depression, the rise and fall of communism and fascism, the atom bomb, and moon landings. What happened on America's farms was little noticed. This decline in the rural population probably is the most fundamental change in American history. As I will argue, its social consequences, such as family breakdown, are being felt mainly in cities. We are finding there is no alternative to the rural base of our urban culture.

Back in 1946, CBS correspondent Eric Sevareid, who spent much of World War II in Europe, wrote in his autobiography, *Not So Wild a Dream,* about what it was like to grow up in the remote little North Dakota farming community of Velva. Sevareid felt the war had ended a little Midwestern town's isolation. He wrote:

> America was involved in the world, all its little Velvas were in the world, and the world was now in them, and neither the world nor America would ever be the same. . . . All that America truly meant, all that Americans perished for, would be devoid of consequence or portent unless the image of society that America showed the world was that of the little Velvas as I had known, remembered and cherished them.

From such communities, Sevareid said, American youths went off to war carrying "America's bright tools and great muscles, her giant voice and will." With the confidence so characteristic of the outset of America's post–World War II imperial era—what Henry Luce hailed at the time as the American Century—Sevareid felt it was this country's destiny to "create a world in its own image." He defined that image in terms of his small Midwestern hometown.

The idealism and optimism Eric Sevareid voiced for his generation in the late 1940s sound strangely outdated today. Who would talk of remaking the world in our image now? It is not just that such sentiments jar a younger, more skeptical post-Vietnam generation or that our respect for other cultures has grown. Our confidence has also waned in the past half century so that it no longer seems natural to offer an American farming community as the ideal cultural model the rest of the world should, or could, follow. Today, wiser from our 45 years of postwar involvement with the world, we recognize that the vast majority of its inhabitants are peasant villagers within very old civilizations—Confucian, Buddhist, Hindu, Malay, Christian, Judaic, Islamic, or tribal African. And ways and customs evolved over centuries from a simple rural economy determine their behavior.

The relative poverty of the non-Western world is partly because its agricultural revolution is coming 50, 60, 70 years later than ours. These years have seen phenomenal advances in radio astronomy, solid-state physics, and molecular biology and such technological advances as space satellites, rockets, transistor radios, computers, compact disks and video-tape, lasers and microchips, jet travel and test-tube babies. And with such advances also have come worsening overpopulation, overuse of resources, and overpollution. Raging fires in the Amazon jungles, like record heat waves on the Great Plains, warn of environmental catastrophe.

As the cold war and threat of nuclear warfare wind down, the spread of Western science and ideas all over the planet is likely to be the chief source of trouble as people succeed or fail to adapt culturally to them. As I reported in the London *Economist* in 1979, "Times change, and men, once they have the technological means and enough years to culturally adjust, change with them." This process is seldom easy. It is particularly difficult for the West itself. Scientific and technological advances, if they move a society too far from its agricultural origins, can set it on a road toward eventual cultural and biological extinction.

The true America, to Eric Sevareid, was the way of life that grew up from the disappearance of the frontier in the 1880s and reached its peak in the 1920s and 1930s. In these 50 or 60 years America went from country to city, farm to factory, horse to car, kitchen surgery to scientific medicine, fundamentalism to Darwinism, tent meeting to Hollywood movie. More subtly, these years also saw a steady weakening of religious belief, family ties, the work ethic, the agricultural moral code, and the small town's sense of community. The breakdown of the American family and social ethics, along with crime, drugs, homelessness, and so on, all go back to our urban society's movement away from its agricultural base. There is simply no substitute for the farm and small town when it comes to forming human culture.

This is not a value judgment. It is a simple anthropological fact. A society can work only as long as a large minority of its people live on and farm the land. This is hard physical work. A great many of us would prefer to live in cities, particularly if we are in a profession. Neither Sevareid nor very many of his generation chose to go back to their Velvas and to the farming and marketing way of life that kept these small farms going. As scientific farming and mechanization advanced—amidst a chronic cost-price squeeze, competition for land, and a patchwork of federal programs that, whatever their intent, consistently awarded bigness—farms got ever bigger and fewer.

As modern technology took more and more of the stoop labor, sweat, and drudgery out of farming, it forced millions off the land. The Okies of John Steinbeck's *The Grapes of Wrath* had to migrate when Dust Bowl drought churned their farms into red dust. The blacks of the Deep South were uprooted when mechanical cotton pickers made them redundant. A huge rural-to-urban migration of Americans took place in the 1940s and 1950s. As whites went from the country to the cities of the North and West, and from there into the suburbs, blacks often got trapped in inner cities, even if they too escaped to the suburbs if they could.

The trend toward ever-smaller numbers of farmers went right on during the 1970s despite—for the first time in 160 years—a higher rural than urban population growth. The 1970s saw agriculture's biggest postwar boom. Exports, helped by a devalued dollar, soared to Russia, Europe and the Third World. But the promise of a reruralization of America proved short-lived, scarcely lasting out the decade. Inflated land values left many farmers with heavy debts when the 1980s brought falling prices and exports, rising farm surpluses, and rising interest rates.

A wave of bankruptcies and farm auctions in the early 1980s was followed by 2 years of drought. In 1988 precipitation on the northern Great Plains fell to less than 7 inches, half the normal amount. A generous $8 billion federal disaster relief program—less expensive than the usual $11 billion in crop supports—plus full grain bins and high prices (durum wheat in North Dakota, for instance, jumped to nearly $7 a bushel, twice the normal price) and federal crop insurance prevented a national calamity.

Most farmers were harder hit in 1989, when there was almost as little rainfall and by then their bins were empty, disaster payments less generous, yields almost as poor, and prices came down. The consensus of farmers I met for this book seems to be they lost about 30 percent of their income both years. And that they will need at least 2 good crop years, with fair yields and fair prices, if they are to get back on track again.

Even with good weather, full recovery won't come until several years into the 1990s, at best. Probably 1986 and 1987 were the worst years. Land values in Iowa, down 63 percent in 1981–87 to less than $800 an acre, were up to $1,000 to $1,200 by mid-1990. In central North Dakota, the other area studied here, land values rose to a peak of $600 to $700 an acre, fell to $350 in 1986, and in 1990 were back to $400 to $450. The

American Bankers Association and the National Agricultural Statistics Service agreed that more than a hundred thousand farm families a year were forced off the land in 1986 and 1987.

In 1987 the Congressional Office of Technology Assessment warned that only 50,000 farms may produce 75 percent of American production by 2000. Just how many farms there are now is open to question. The Department of Agriculture classifies any land holding that makes at least $1,000 a year as a "farm." Officially there are 2.2 million U.S. farms by that definition. Some authorities, such as the University of Minnesota's Professor Vernon Ruttan, say a more realistic figure would be 250,000. Others dispute this as too low.

How do we keep the number of farms from falling still more? In the conclusion I'll give suggested solutions from a range of experts and offer my own. Here let me just say that farm policy has always been confused by the vague but firm conviction among Americans that life in the country is somehow better and purer than life in the city. As I'll argue, we have good, practical grounds for feeling so. Small farmers, and to some degree those who own small-town businesses, are such folk heroes that all Democratic and most Republican administrations have found it politic to commit themselves to keeping them going. Many farmers told me the 1988 drought disaster relief was so generous because it was an election year. I think television coverage of stricken farms and small towns genuinely scared a lot of ordinary Americans. At the same time, there have been big and continuing cuts in crop supports, as in the 1990 farm bill.

The biggest reason farm ownership has steadily become more concentrated has been the use of bigger and bigger machines. Chemistry, in the form of pesticides, herbicides, fungicides, and fertilizer, has played a role, as has plant breeding. But the speediest growth in productivity has come from tractors and combines. Travel the wheat belt of the Great Plains, and the fields are empty. Machines as big as brontosauruses have replaced the old tractors and people. The new machines carry very fancy price tags. A 135-horsepower tractor easily cost $50,000 to $60,000 in 1990; a combine harvester, $125,000 to $135,000; and an oxygen-free silo, over $50,000. Center-pivot irrigation systems, mechanical harvesters, and infrared aerial photography, crop-dusting planes, and computerized electronics for monitoring crop conditions also come expensive.

Does this mean that the bigger farms get, the better they are? No. Practically everybody agrees that a medium-sized farm, the definition of which varies with geography and crops—about 800 to 1,000 acres in wheat-growing North Dakota, half that for the corn and soybeans of Iowa—is usually more efficient than a farm that is either too small or too large.

Why? If a farm is too small, its efficiency declines relative to that of larger, more efficient neighbors who can afford investments in advanced techniques and machinery. Prices get set by the bigger operations, and the small farm can't afford the new machinery it needs to survive. For years just able to eke out a living, such small farmers are pushed out of farming. They have no trouble selling land. The big farmers are always looking for extra land so their new equipment is fully employed.

The idea that faceless corporations are taking over American farming is a myth. There *is* a danger they will try to control inputs or markets. But the consolidation of American farms themselves into bigger units is due mainly not to giant corporations investing in farming—illegal in many states—but to the expansion of family farms themselves. Big family farms swallow up little family farms. Yet even the trend to huge farms and its concentration of ownership, often shutting the door to young people who lack family advantages, itself seems to have its limits. One often-heard explanation is that a farm family has more incentive to work the 16-hour days needed to get the crop put in or taken out, or it can be idle without pay in winter or when it rains. Efficiency declines with hired, less-motivated workers.

In the late 1980s land consolidation was slowed down by economic crisis and drought. Bigger farmers no longer bought everything in sight. Some went broke. But consolidation is likely to continue, if at a slower pace. Congressional efforts to protect the smaller farmer through changes in the tax system have not worked; the biggest tax savings go to the biggest farmers.

A radical solution would be to impose a land ceiling and redistribute land from large to small farmers. No constituency, however, yet exists in this country either for redistributing land or putting limits on its acquisition. Certainly not among farmers. Ask any of them about it; the idea strikes them as crazy. The radicalism of American farmers extends only to their adaptability toward new techniques and machines. Politically and culturally the farmer is a conservative, with an abiding respect for the traditional way of doing things.

This conservatism has produced the main traditions of America's post-frontier rural society: small, church-going families, authoritative fathers, moralizing mothers, and children who grow self-reliant by performing useful chores from toddlerhood. Farmers and their families cling to these values far more tenaciously than city people. Such "save the farm" films as *Country, The River, Places in the Heart,* and 1989's *Miles from Home* show that at least nostalgia for these traditions is fairly widespread among all Americans. Rural ways still have such box office appeal in the city because the countryside is where our deepest cultural feelings lie.

It is this contradictory cultural divide—making the most of new farm technology while holding on to the old values—that is the American farmer's real strength. In just 130 years farming in the United States has gone through two revolutions. The first, starting from the Civil War, was from human power to horsepower; the second was from horsepower to mechanical power (the number of tractors did not exceed the number of horses until 1955). Along with mechanization came much greater application of biology and chemistry to farming. This meant better seeds and better breeding, and more sophisticated, if now increasingly controversial, use of chemicals as fertilizer and insect and weed killers.

Much as we try to look at things as they really are, we are all influenced by our own historical and geographical backgrounds. At the outset I should mention that I am a middle-class, middle-aged (59) Midwesterner of semirural origins. One grandfather (1854–1926), a New England Quaker, helped break the sod as a settler in Iowa, grew up on a farm, and spent most of his life as a country doctor and later a rural pastor. The other grandfather (1861–1904), descended from Virginians who migrated to Ohio, was a pioneer country doctor in North Dakota's Red River Valley. My father (1889–1938), after farming 7 years, was a country doctor too. My mother (1887–1982) rafted down the Mississippi in 1904, taught in a one-room schoolhouse, went off to North Dakota seeking adventure, farmed herself when everybody got influenza in 1918, and widowed early, educated five children and lived to be nearly 95.

When I was doing research for a social history of America from the 1880s to 1940s, drawing on my own family's experience, I came to feel during many stays in Iowa and North Dakota in the early 1980s that whatever ailed urban America, the key was somehow to be found out here where the country was at its most rural. Going back to the Midwest later in the decade, I decided to focus on the two communities where I did most of the research for the earlier work. The first, where I was born and which I'll call Prairie here (population 761), is so remote out on the sparsely populated North Dakota prairie, its people can survive only by farming. The second, Crow Creek, in Iowa—also not its real name—(population 369) has gone from an old-fashioned farming community to a virtual suburb of the city of Cedar Rapids, just 25 miles away. Crow Creek is now settled, except for a few old people and its outlying farms, almost entirely by commuters, many of whom work for a high-tech radio and avionics manufacturer which outfits even spaceships. To me, Crow Creek's people exemplify the alternative fate for rural Americans who stay on the land: if they cannot make it in farming, as Prairie's people must, and industrial and service jobs are within commuting distance, a community like Crow Creek can still survive. Unlike Prairie, the town itself does not move toward extinction. Only its rural culture does.

What do I mean by culture? As everybody who watches television or reads a daily paper knows, we Americans have a habit of looking at problems in terms of the politics of the surface, instead of the economic and cultural trends beneath the surface. This makes us slow to recognize long-term economic and cultural changes which require changes in policy. In this book I'd like to invite readers to view rural America in an essentially anthropological way. Cultural adaptation and ways for the small family farm and rural community to survive are what we should be looking for and trying, if possible, mainly economically, to assist. The way to solve the farm problem, if there is a way, is on the farm. But the culture we are really trying to save by keeping enough people on the farm, make no mistake about it, is the increasingly secular, scientific, anonymous, unstable, highly differentiated culture of urban America. It is our own.

The same is true of the Third World. Our foreign aid priority ought to be to try to find economic means to keep its vast rural populations in their villages. The Third World's hope is that the oldest human cultures can adapt to the newest scientific knowledge within the context of a village-based agrarian civilization. As I mentioned, they are not going to become Velva-like communities, but something, in cultural terms, distinctively their own. As I put it in *Foreign Affairs* in 1982: "The villagers are not moving from their (A) to our (B). Rather we are all moving toward (C), a wholly new society based upon biotechnology, electronics, new energy drawn from water and sun, and all the other new scientific advances." Whether they, or we, achieve this new society or not depends on keeping enough of them—and us—in small farm communities. If this does not happen, I'm afraid our cities, if very, very slowly, are moving toward cultural extinction. The social ills we face now will get steadily worse.

Until the 1980s I knew next to nothing about American farming, though I'd spent some summers working in an uncle's fields as a boy. I'd spent much of my adult life—from my first trips to Korea in 1954 and India in 1959 to my most recent long village study, in Kenya in 1981, and journeys to the famine-stricken African Sahel in 1984 and China and Indonesia in 1985—observing and writing about how everyday people in the Third World, mainly villagers, handled change. As I have suggested, this experience is not so unrelated to what is happening in rural America as it might at first glance seem.

My essential finding in villages in Asia, Africa, and Latin America over all these years is that village life is not only vital in itself but is also the fundamental basis of all civilized behavior, including our own. Third World villagers and the people of communities like Prairie and Crow Creek exist in the same continuum stretching unbroken through time. They all seek an economic basis to allow groups of people to live comfortably, productively, and freely on the land. Americans stand at one end, the most technologically advanced society ever. Yet, as we shall see, almost everybody interviewed for this book is struggling to reconcile the effect of technological advance—ever bigger, fewer farms—with emotions and values formed deep in our past.

Years of experience have convinced me that the same yardsticks used to measure the way people adapt to change in Third World villages apply to American farm communities too (and even, as I'll argue in a minute, to contemporary subcultures like that of America's blacks, or even to whole societies like Great Britain). I use *culture* in the modern anthropological sense to mean a set of rules or solutions to problems, handed down from father to son (or more often in infancy and childhood from mother to son and daughter) so that each new generation does not have to start out from scratch.

First of all, culture is a kind of ready-made design for living. It is inherited. Culture differs from a "lifestyle," which is something individuals themselves choose. The tension between freely chosen individual lifestyles and inherited culture-based group living is pronounced in America today, particularly in urban settings. Freedom of choice in excess can lead to self-enrichment and self-gratification at the expense of the kind of civic spirit and self-sacrifice for the sake of social solidarity that keeps the life of a group going. Groups can be freely chosen (such as a hippie commune in the 1960s) or politically imposed (a state-run collective farm), but these lack the cultural foundation of a village and tend not to last very long in historical terms.

Second, culture has an economic basis. Richard Leakey, who has traced human origins back 2 million years, argues that people are not innately anything (peaceful, warlike, good, evil) but are capable of anything. What decides how they act is culture, based on how they get their living. Leakey says, "Human beings are cultural animals and each one is the product of his particular culture." Originally, he told me in a 1981 talk in Nairobi, humans evolved a cooperative society because the hunting and collecting that kept them alive required it. Agriculture, once it was invented, also depended on a cooperative society and it persisted. (Indeed, the you-help-me-get-my-harvest-in-and-I'll-help-you-with-yours ethic gave rise to the Golden Rule.)

Third, an observation of Aristotle, from his study of humans and animals 2,300 years ago, takes us another step: he found that every creature's mode of living, that is, culture, is decided by how it obtains its food, whether as hunter, collector, herder, cultivator, or predator upon its fellows. If you change the way people get their food, you are going to change their culture too. A stagnant or unchanging agriculture, as, until the mid-twentieth century, in India or China, creates a traditional peasant society with deeply entrenched customs. (Hence, Mao's famous remark to Nixon that he had only been able to change "a few places in the vicinity of Beijing.")

Fourth, all small farming communities have something of a universal culture. They are all groups of people tilling land. They naturally put prime value on property (land, the means of production) and family (tillers, the basic work unit). This is as true of an Iowa or North Dakota farmer as a Punjabi Sikh or Javanese rice peasant. Property and family, along with religion, are man's most basic institutions.

Aside from race or level of technology, I find societies differ most when it comes to the realm of abstract ideas, or religion. Agriculture-based cultures the world over differ most strikingly from one another in the inherited attempts by their early ancestors to provide life and the world of their times with metaphysical meaning. Religion is the core of any culture. It is why a society reacts to the very rapid spread of Western ideas and science with relative ease or difficulty. Today, the Confucians are roaring ahead, followed more hesitatingly by the Malay-Javanese and Hindus. But all three purely Asian cultures are adapting more easily, in the Third World, than Muslims, Christians, or tribal Africans.

Fifth and last, since science—technology being its practical application—is the agent of change and religion is the core of any culture, when we observe changes in a rural American community, somewhere in the picture we can expect to find some kind of confrontation, though not necessarily an outright conflict, between science and religion. I found this much more marked when doing research into the 1880–1940 period, suggesting science has won most of the battles by now, at least for the time being.

I've mentioned that these yardsticks can be used to measure, compare, judge, or explore any society. Take America's blacks. They have been culturally uprooted twice, first from tribal African beliefs based on a hunting, herding, hoe-cultivation economy (where women were often the farmers). And second, after the invention of the mechanical cotton picker, from rural communities of the American South.

Some cultural traits, such as the mother-centered family or the African abstraction of a personal God, have been passed down from generation to generation, miraculously surviving today. Many blacks move into the middle class and assume middle-class cultural values. But others are trapped in crime- and drug-ridden inner-city slums where a black underclass is emerging that hardly knows any longer what the rules are. Cut off from its roots both in Africa and the rural South, this underclass is losing its old inherited design for living. Fully 95 percent of all black share-cropped farms—about 350,000—disappeared between 1950 and 1970, with hardly a murmur of concern from anybody. The already small black farm population was cut by half again in the 1980s, from 243,000 to 123,000. The Congressional Office of Technology Assessment warned in 1988 that of 33,250 farms found to be still owned by blacks in 1982, almost all will be lost by the mid-1990s.

This is *culturally,* not just economically, a dangerous development and should not be allowed to happen. The urban riots of the 1960s have not recurred, even though the social isolation of inner-city blacks is worse than ever and so is youth unemployment. What has happened instead is a huge increase in destructive behavior by individuals: vandalism, theft, violence, muggings, school dropouts, drug abuse. Nearly a quarter of young American blacks are now in prison; more are in jail than in college. Blacks, who make up 12 percent of 248 million Americans, are the first large minority to be so totally cut off from the rural base of their culture.

In *An American Looks at Britain,* I used these cultural yardsticks to try to comprehend why there has been such a wholesale collapse of invention and enterprise in that society. Mrs. Thatcher exhorts the British to go from "a dependency culture" to "an enterprise culture." But the cultural decay goes deep. First, to look at the economic basis of British culture. In 1800, when most Europeans were still subsistence peasants, Britain pioneered commercial farming. It introduced new crops, horse-drawn machinery, and new irrigation and drainage systems. These meant only a third as many people as before were needed to farm. This drop in the needed rural work force coincided with the invention of the steam engine. Quite suddenly Britain, which already had navigable rivers, good harbors, and networks of roads and canals, found itself with the technology, farming base, and labor force to launch the world's first industrial revolution. This in turn thrust Britain into its Empire, seeking trade.

The population tripled in the nineteenth century and poured into the new mines, ironworks, and textile factories; Britain today is 80 to 90 percent urban. With the twentieth century, two world wars, and loss of empire came not only industrial decline, but the most spectacular collapse of organized religion yet seen anywhere. The Church of England, in a nation of 57 million people, can no longer count on getting more than 1.5 million people to church on any Sunday; just about as many Muslims go to mosques in Britain. When I interviewed Anglican church leaders about it, they spoke of Britain as "a secular country" and "a postreligious society" with a "breakdown in values." Ronald Bowlby, the Bishop of Southwark, or the southern half of London, said, "There is something hollow about our society. There's something wrong. There's something missing. There's no heart in our society anymore."

David Jenkins, the Bishop of Durham, blames Britain's religious decline on the postindustrial electronic age and says it is "on the leading edge of the breakup of the old communities. . . . We have now stopped being a Christian country."

How does this relate to the rural issue? Very directly. Listen to the Bishop of Southwark's explanation: "All the people who flooded into the English cities during the nineteenth century were rural. They had an ingrained kind of religiousness. It wasn't particularly formed or church-going. But it was there. And it is that which has finally drained away. People say the Church ought to work harder to reclaim the working classes. The truth is it never had them."

This is just as true in America. Our rural exodus is even more recent. In 1920 when farm residence was first added to the census, 32 million Americans said they were living on farms. That was 30 percent of the total U.S. population then of 106 million. Calvin Beale, the chief demographer of the Department of Agriculture, says, "It is believed that the farm population peaked at about 32.5 million in 1916, and was already beginning to decrease when first counted in 1920. Essentially all cultivatable land had been occupied, mechanization was underway, and the employment boom of World War I pulled people into the cities. People on farms gradually declined to 30.5 million by 1930, the outset of the Great Depression. The Depression decade did not see any further decrease in total farm people as the losses from the Dust Bowl were offset by gains in other regions where people remained on small farms as a refuge from unemployment." Hence the peak year for the number of American farms, 6.8 million, was in 1935; for people on farms, as Beale says, it was 19 years earlier.

With World War II, the farm population dropped 6 million people in 1940–45, the fastest drop ever. By 1960, the number of farm people was down to 15.6 million, barely half as many as in 1950. Which means that if about a third of Americans over 70 have a first-generation rural cultural base, only a small number of those 50 and under do. The generations directly shaped by rural life are dying out.

If one can condense human history into a single paragraph, we might say that something like a man—*Australopithicus*—hunted and collected for 2 or 3 million years in the sub-Saharan savannah where it evolved, cut off from the rest of Ice Age earth. By a million years ago, *Homo erectus,* who had learned to make fire, shape tools, and probably talk, broke through the Saharan barrier, presumably by moving down the Nile Valley. Humanlike bones have been found in Europe, Java, and North America (the Americas weren't reached until *Homo sapiens*). Neolithic woman, as the collector of berries, roots, and wild grain, invented agriculture 10,000 to 15,000 years ago. The plow, along with irrigation, led to the rise of civilization in the Fertile Crescent 6,000 years ago. The invention of the heavy moldboard plow and the dryland manorial farming it made possible led to the rise of Europe, starting about 1,500 years ago. This was followed by the industrial revolution 200 years ago and the West's technological domination to this day. It is only in the late twentieth century that the Asians began to catch up.

For the West, 1800 was the key date. From the day men followed their cattle down from central Asia and up from Africa into the Fertile Crescent, and invented the plow, wheel, sail, and irrigation, getting settled cultivation on its way, nothing really big happened to the farmer—aside from the invention of the moldboard plow and gunpowder—until about 1800. There were great nonfarming advances during these 5,000 or 7,000 years, such as the Gutenberg printing press and techniques of seafaring and navigation. But farming, the world over, stayed pretty much the same. The best account of agriculture's role in history, which shows that major social changes are usually triggered by new foreign technology, is William H. McNeill's *The Rise of the West*. Arnold Toynbee called it "the most lucid presentation of world history in narrative form I know," and H. R. Trevor-Roper called it "the most learned and most intelligent" history of mankind yet written. Professor McNeill, when I visited him at his country home in Connecticut, warned me not to simplify. One cannot overlook other changes in farming, he said, giving an example: "The rice paddy was perfected not before 1000 B.C. in all probability and American food crops transformed world agriculture after 1500."

But very generally speaking, I think it is safe to say that anyone cultivating land in A.D. 1800, whether in Europe, China, India, or what was then the 24-year-old United States of America, used about the same energy sources as the 5000 B.C. cultivator (animal power, wind, water, sun, and his own muscles). He could travel much the same tiny maximum distance per day (walking or sailing or by horse or carriage—the first railroads date from 1820). He used much the same material for tools (iron and wood) and fuel (firewood and forage for draft animals). He had much the same life expectancy from birth (to his late 30s or early 40s, about what people have today in Chad). And then in the West after 1800, everything suddenly went whoosh.

As Professor McNeill shows, new technology can quickly shatter the stability of a society. One sees so many examples of it. I remember when I visited Iran's Khuzestan Desert in 1974, French archeologists digging on what was once the eastern edge of the Mesopotamian Plain said the first known villages there dated from 8000–4000 B.C. They gave me charred wheat kernels from one of the oldest, 10,000-year-old sites. (I put them in a matchbox where they soon crumbled to dust.) For about 4,000 years these villages survived, as far as archeologists can tell, in a state of peaceful anarchy. Irrigation, thought to have been first invented here, shattered this calm very suddenly. It led, they say, in a relatively short space of time, to a surplus food supply, the development of the first towns and cities, a rapid growth in population, and a decline in the absolute number of villages. Intercity warfare began almost with the first Mesopotamian temple communities, along with the construction of defensive walls, the abandonment of outlying villages, migration to ever-larger urban centers, and the rise of soldiers, armies, generals, kings, and sovereign states like Elam and Sumer and Akkad. (I went to this area because of a map in the paperback edition of *The Rise of the West,* showing ancient Mesopotamia extending from then unfriendly Iraq into then friendly Iran.)

Norman Borlaug, the Iowa plant breeder who won the 1970 Nobel Peace Prize for his work in dwarf wheat—what really launched the Green Revolution—and who has been my best source on agricultural science over the years (his father farmed and my grandfather preached in the same part of northeastern Iowa), says another good example is Jericho. A food surplus from wheat made possible this first-known city 9,000 years ago. But once the cultivators started to store grain, nomadic shepherds from the desert attacked and seized it. To fortify themselves, the people of Jericho built a huge watchtower and the famous walls which came tumbling down in Joshua's battle. (In 1990 archeologists reported new evidence that the walls did in fact fall at the right time, though it is deemed more probable from an earthquake.)

Here too, as in Mesopotamia, small family-sized farms of free men were replaced by bigger estates, farmed with economies of scale by serfs and later by slaves captured in war. Like the first temple cities, some of history's brightest civilizations—pharaonic Egypt, China and Japan under the seventh century T'ang dynasty and Fujiwara clan, Mexico under the Mayas and Aztecs—practiced something close to modern state socialism, with heavy taxation of the food-producing peasantry.

In Europe, the disintegration of traditional peasant society began with the introduction of the heavy moldboard plow. It was hastened when calculations of price and profit in the medieval cities began to modify crop rotation and methods of cultivation. The death blow to peasant agriculture came in the eighteenth and nineteenth centuries with the gradual spread of modern farm technology—based on new findings of science—and the treatment of land, labor, and rent as commercially negotiable properties. As mounting debt blanketed the villages of Europe, fifty million peasants migrated to North America in the century after 1800.

At the end of the Revolutionary War, America had just 2.8 million people, 61 percent of them of English descent and culture (82 percent in Massachusetts). Our first census that year showed 96 percent of Americans were rural. (In the world today, 60 percent are rural—in Asia 73 percent, in Africa 72 percent, in Latin America 35 percent; and in the industrialized West 29 percent, Britain a mere 10 percent.) Most early Americans lived on farms; they turned the earth with wooden plows, sowed by hand, mowed hay and grain with sickles or cradle scythes, and threshed with oxen and horses (all still to be found in the Third World today).

In 1791 Thomas Jefferson made his famous remark that "those who labor in the earth are the chosen people of God, if ever He had a chosen people." But Jefferson was just as fascinated by new technology as any contemporary farmer. He designed an improved seed drill, a brake for separating the fibers of hemp, a threshing machine, and an improved moldboard, that part of a plow blade that turns the soil. George Washington told Congress in 1796 that "it will not be doubted that with reference either to individual or national welfare, agriculture is of primary importance. . . ." Washington didn't invent his own machinery. He did ask Arthur Young, an English advocate of agricultural advance, to send him the latest farm implements from London. In Jefferson and Washington we see, right from the start, the American paradox of trying to reduce the need for human muscle in farming while lamenting the falling number of farmers.

The best-known new farm technology just after the revolution was Eli Whitney's invention of the cotton gin in 1793. Upland cotton, the kind grown in the South, has fibers that cling to the seed. Whitney's gin gave farmers a way to separate the lint from the seeds. Cotton production rose in consequence from about 10,500 bales in 1793 to nearly 4.5 million in 1861, as plantations grew and became reliant on slave labor.

In New England the availability of this low-cost cotton and of new spinning and weaving machinery from England led to a rapid industrialization. Quaker merchant and philanthropist, Moses Brown—he helped to found Brown University, named after his brother—and Samuel Slater introduced the new machines. (Brown's son Obediah married my great-great-aunt Dorcas.) New England's surplus food supply provided cheap fare to the mill workers. Young people left farms in droves for jobs in the mill towns, working for low wages.

Old letters handed down for several generations in my family are full of farming details. One can learn that 50 bushels of wheat, and the same with oats, were threshed from "little more than an acre" in 1838 in Jacksonville (now Ithaca) in what was then the frontier in New York state. Five hogs weighing a total of 1,442 pounds sold for $95; 180 pounds of sheep wool went for 54½ cents per pound, a poor price as "it was washed very badly." A mower earned $20 to $28 a month cutting grain with a cradle scythe; one is described by my great-grandfather's sister as "the handsomest mower we ever saw." Threshing was done with flails or by oxen or horses who trod out the grain from stalks spread out on the threshing floor. Flailing was considered the worst back-breaking, arm-aching drudgery on a farm; a common sight was a ring of threshers beating grain with flails on a straw-littered barn floor. To winnow, grain was tossed into the air so that the chaff was blown away. They kept the barn doors open as there were great clouds of dust.

One batch of letters, received by my grandfather in Jacksonville from relatives in Watertown, New York, and Providence, Rhode Island, his original home, all dated 1835–48, gives a graphic picture of early American rural life. Education was by a tutor, who charged a tuition. A stagecoach picked up passengers at "the tavern in the Hollow." Life was not crime free. One letter told how a village store was robbed of $400, a fortune in those days. A whole gang of local Quaker youths were involved, the sons of respectable farmers. They were kept in irons, for "there is such a company of blacklegs as they call them that they are afraid they will rescue them." The language in these letters is strikingly English, a quality that gets lost as soon as their authors migrate to the Middle West.

Early nineteenth-century American farming communities had many cultural traits in common with present-day Third World villages. This is not surprising, of course, as their economic bases are so similar. Generally people who farm the world over are plain, straight, and conservative with a present-time orientation and concrete mindedness. In the Third World, as in nineteenth-century rural America, one finds the same respect for age and custom. A father has to provide food, shelter, and clothing for all the family members. And each member in turn is obligated to work under his direction; the father's authority has an economic basis. The family is of central importance. Blood ties and kinship have heavy weight.

Early Americans, like today's peasant villagers, had the same reverence for nature, the same desire to own land. Hard physical labor was an unchanging fact of farm life. It was a habit inculcated from childhood. A boy was as prepared to farm at 15 as he was at 40. Industry and thrift were prime values. Monogamous, divorceless, and multichild marriages were the rule. Many children were welcomed as more hands for work and more security in old age. Neighbors depended on mutual help. Life was governed by the seasons. Social life revolved around births, marriages, deaths, and going to church.

Life expectancy in America in 1800 was just about what it is in the poorest Third World countries today. Even by 1900 rural Americans lived just an average 47 years, 18 of them in the child-bearing age. This was what it was in Bangladesh last year.

Some of these traits have utterly lost ground in the rural America of the 1990s. Others vigorously survive.

The village model of my New England ancestors—a cluster of houses from which people went out to farm—was not repeated in Iowa. Instead, like other westernizing New Englanders or Southerners or uprooted European peasants who settled in the Great Plains, those in Iowa built in the German style of scattered, individual farms surrounding a small trading town. At first, as in Crow Creek, farm settlements were tied to rivers and streams, which were, with so few roads, the main routes of transportation. Later, as in Prairie, settled 50 years afterward, they were tied to railroads.

Most of these new communities were villages really, but right from the start their inhabitants called them small towns. (Villages, both as a word and a reality, survive only in New England.) Most of these Great Plains pioneer towns were dominated by a Yankee commercial and professional stock which actively discouraged old European customs of what was often an immigrant farming majority. For their part the immigrants, most of them ex–European peasants who had cultivated their land largely for subsistence and to pay a landlord, now found that they had to become commercial farmers in a business for profit if they were to survive. It was a hard school and its story is well told in such works as Willa Cather's *O Pioneers!* and *My Antonia,* Oscar Handlin's *The Uprooted,* or O. E. Rolvaag's *Giants in the Earth.* This cultural adaptation shaped a distinctly American farm mentality: independent, adaptive, inventive, entrepreneurial, ready to try every new gadget, but ultimately deeply conservative.

Abundant land and a sparse but growing population partly explain why American farming became so commercial so soon. But the main reason, I think, was the post-1800 explosion of American farm technology.

When Cyrus McCormick built his first reaper in 1831, harvesting was still universally done by hand. Cast-iron plows had been patented as early as 1797, and some farmers began to use them, though there were claims iron poisoned the soil and made the weeds grow. Neither wood nor cast-iron plows worked well in the sticky black prairie soil of the Middle West. Earth stuck to the plow instead of sliding by and turning over. In 1833 John Lane, a blacksmith from Lockport, Illinois, fastened strips of steel meant for saw blades over wooden moldboards. His plow worked, but he didn't patent it.

Then 4 years later, in 1837, a blacksmith whose name is close to being synonymous with farm machinery today, John Deere, began making plows out of saw steel and smooth wrought iron. The plows were popular and Deere went into the business of making them. Needless to say, the company is still going strong; it is Iowa's biggest employer.

The reaper came in 1831 and the steel plow in 1833–1837. The years

1830–50 saw a whole rush of other technological innovations in farming. The combine was invented in 1836, the threshing machine in 1837, the two-row corn planter in 1839, the grain drill in 1840, the grain elevator in 1842, and the portable steam engine in 1850. Corn shellers, hay-baling presses, and cultivators were also devised.

What mattered most was McCormick's reaper. Harvesting, the gathering-in of a crop, is the most important work in farming. Pliny the Elder tells how an oxen-pushed reaper was used by the ancient Romans in Gaul. A comb that projected from the front of a box on two wheels tore off the heads of the grain, which fell into the box. Even so, Americans were still cutting grain with a cradle scythe in the first half of the nineteenth century. In India's Punjab today, though most grain is threshed by tractor-powered machines, much of the harvesting itself is still done with sickles. As I described it in *Villages:* "We cut the wheat as the Punjabis always had, the men crouched low on their haunches in a wide, spread-out line, slashing sickles at the dry stems, grasping a handful at a time, advancing slowly, rocking from side to side, moving steadily and rhythmically down the field."

The first patent for a reaper was issued in England in 1799. A horse-drawn reaper was first patented in this country by Obed Hussey in Maryland in 1833; he set up a factory in Baltimore. Cyrus McCormick patented his 1831 machine in Virginia in 1834. Within 20 years, McCormick led the field, partly because he moved his manufacturing plant to the heart of the Great Plains in Chicago. Together with John Deere's steel plow, the McCormick reaper helped make possible the Great Plains' rapid settlement; it could do the work of forty men with sickles. By 1851 McCormick's Chicago plant was turning out over a thousand reapers a year. As textiles had done in New England, farm machinery was leading to the industrialization of the Middle West.

Europeans and Americans, unlike many Third World people today, particularly Muslims, were well prepared culturally for the post-1800 explosion of technology by the sixteenth-century Protestant Reformation. The Protestant ethic, as Max Weber termed it, provided a psychological underpinning by transforming Christianity into a rational, ethical, world-affirming religion. Reformers like John Calvin assured an urban middle class that Christians could seek salvation through hard work and a more scientific control over matter and energy. Also, by breaking the hold of the church upon life, refusing to take authority for granted, the Reformation influenced the emergence of such religious movements as Quakerism, which had no clergy at all. This freedom of dissent was slowly to produce a secularization of life and attitudes among all Americans. Among Christians, it did not much extend to the Latin Roman Catholic world.

Robert S. Lynd and Helen Merrell Lynd said in *Middletown,* their 1929 study of American culture based upon research in Muncie, Indiana, that they intentionally chose the Middle West as exemplary. The Lynds found Americans in Muncie did six main things: they made a living, kept a home, trained the young, used leisure time, engaged in community life, and practiced religion, the activities of most people the world over.

The Lynds also quoted J. Russell Smith's observation in *North America* that the two main streams of colonists met in this region of the United States; that is, the Yankees of the north came into northern Ohio, while the southerners traveled into Kentucky and down into the Ohio River.

While writing *Those Days,* the social history earlier mentioned, I found both geographical and cultural strains of early American—Yankee Puritan and easy-going Southerner—as well as Scots and Irish immigrants—had gone to make up my own family. From the Yankee Puritan strain came a great-grandfather who migrated with his family in 1859. Tent meetings and evangelical revivals were sweeping rural New York state, and when some members of the Society of Friends turned to preachers, hymn singing, scripture reading, and sermons, he joined a Fundamentalist Quaker faction which broke off to protest what they considered dangerous departures from orthodoxy. They founded an old-style Quaker village in Iowa, though I suspect, like many New Englanders, my great-grandfather mixed religious zeal with a shrewd appreciation of the prosperity to be gained by clearing and farming virgin soil.

The rural southern strain went back to a Welsh convict who, impressed as a seaman on a British frigate, escaped to the backwoods of Virginia. His sons, barely literate and common foot soldiers in the Revolutionary War, got "Ohio fever" and took their families to clear a tract of southern Ohio bottomland. They closely fit a description by Richard Lingeman in his *Small Town America* of the southern stream of pioneers who wanted little but land to farm and to be left alone. He wrote:

> They were independent, beholden to no man, suspicious of the law, indifferent to schooling, hair-triggered at fancied slights to their honor; they were also tough, hardy, crack shots, gifted at surviving in the wilderness. The Southerners were generally less religious than

the New Englanders, though prone to revival-meeting purgation; they were less likely to link public morality and religion than the New Englanders and were thus more tolerant of the kinds of vice the New Englanders were always trying to legislate out of existence with blue laws.

According to Lingeman, Yankees tended to be better educated and generally went west with more money, from the sale of farms back east. (My great-grandfather kept his in gold, hidden at home.) The Southerners "came from a society that relegated the poor to ignorance." Yankees were known for sharp practice and thought the Southerners "irreligious, immoral, lazy and dissolute." Lingeman observes that the Middle West became an amalgam of these strains, the Southerners bringing to frontier life "their hospitality and generosity to strangers, as well as a love of liberty and an ingrained distrust of authority" and the Yankees "their abiding respect for education, their enterprise, their puritanical morality, and their township system of settlement."

In my own family, the Yankees' descendants stayed on in Iowa, mainly as farmers, with a few teachers, preachers, and doctors among them. The Southerners' descendants, after a few generations as poor backwoods farmers in Ohio, became lawyers, got rich, and had an addiction to fast horses. The grandfather who became a doctor in North Dakota arrived in 1886, just 10 years after Custer's "last stand" at Little Big Horn, and he may have hoped to get rich quick through wheat and land speculation, now that the Indians were vanquished. At any rate, he arrived with fifteen trotters from Ohio, built a forty-room Victorian mansion with hot and cold running water out on the prairie at a time most settlers lived in sod houses, was famed for driving "the fastest horses in the county," and died in his early 40s after a buggy accident.

The 1830–50 rush of new farm technology was well publicized, but many farmers were hesitant to invest in so much new horse-drawn machinery, especially when hired hands were cheap and abundant. It was not until the Civil War (1861–65) that so many farmers shifted from hand power to horsepower. The years 1850–80 are sometimes called the first American agricultural revolution. The other two quantum leaps forward for mechanical technology came during and just after World War I (1914–18) and World War II (1939–45), in all three cases in response to manpower shortages and high farm prices.

In 1850–80, though average crop yields stayed the same, there was a big jump in worker productivity. This set America on a path it still follows today: as the University of Minnesota's Professor Vernon Ruttan has pointed out, the biggest gains in production have come from mechanical technology, or advances in output per worker. First from horse-powered, then from oil-powered, machinery. This has not been the pattern all over the world. In Japan and parts of Europe and much of the Third World, the biggest gains in farm production have come from biological technology, advances in output per unit of land. As our knowledge of molecular biology expands, as we learn more about the toxicity of chemicals, and as oil gets scarcer and more expensive, this is the way we will possibly go too.

As America started down its machine-based farming road a century ago, pioneers pushed steadily westward. In 1800 a vast area between the Appalachians and the Mississippi lay open to settlement, and in 1803, with the Louisiana Purchase, America once again nearly doubled in size. By 1850, either through treaty or by war, Florida, Texas, Oregon, and the desert Southwest between Texas and the Pacific had been brought into the country's boundaries. It was a vast amount of virgin land.

America now had over 2 billion acres. Indian claims to land were seldom recognized, as we will see is still true today, and southern blacks were still slaves, but under an 1820 law almost any white settler could buy an 80-acre farm directly from the government for as little as $100. By 1850, there were one-and-a-half million farms.

The new farm machinery, the end of Indian attacks, federal land policies, good virgin soil, the coming of railroads, growing demand for American farm products at home and abroad, and improved farming methods led to a sudden surge in the number of American farms. They went up by nearly 600,000 to 2,044,077 in just 10 years, 1850–60. In the 1860s, close to 1.4 million more new farms were established, and nearly that many in the 1890s. By 1910 the number of farms was close to 6.4 million. As mentioned, it peaked at 6.8 million in 1935 and has since dropped to 2.2 million. Since, as I said, any rural landholding that raises and sells produce worth at least $1,000 a year is officially deemed a "farm," what you and I call farms would be much fewer. About a third of Americans lived on farms by the turn of the century. The entire rural population, including small towns of less than 2,500, went from about 18 million in 1860 to 35.8 million in 1880 and to just under 50 million by 1910. If we take just those on farms, the total, as I mentioned, is believed to have peaked at about 32.5 million in 1916.

Even then urban America was already growing a lot faster. From 1860 to 1910, farmers dropped from two-thirds to one-third of the labor force. A tremendous food surplus made this possible. Wheat production alone, much of it harvested mechanically, multiplied sixfold in 1850–1900, though this intensive machine cropping was to help produce the disastrous Dust Bowl of the 1930s. But in 1800 farming produced about 40 percent of Americans' private income; by 1900 this was down to 20 percent. Nor could farm income keep up with city wages. In 1880 income per worker in agriculture was $252 a year, compared with $572 for nonfarm workers. By 1900 the gap was bigger: $260 on the farm and $662 in the city.

Some economists make much of this gap, but the University of Chicago's Theodore Schultz, who won the Nobel Prize in economic science in 1979, told me to use caution on these figures. He said, "When the income in agriculture is compared with nonfarm workers, the difference is adverse to agriculture, more than two to one. This is an illusion. Real incomes were much closer. If you standardize by age and experience, they were closer still." If you figure in a farm worker's free meals, free place to sleep in the bunkhouse, free wagon and buggy rides in and out of town, and on the average, the worker's relative youth, this is true, though we will hear from other economists who argue that the comparatively higher urban incomes provided the main imperative to mechanize farming.

Three seminal pieces of legislation were passed by Congress in 1862: the Homestead Act offered 160 acres to any qualified citizen who would live on it for a specified time; the Bureau (later Department) of Agriculture was established; and the Morrill Act set up land-grant colleges by giving 30,000 acres of public land for each senator and representative to provide for an agricultural and mechanical college. Dr. Schultz also cautioned me about the Homestead Act: "It was never really free land; there was a fee. A settler had to settle on it and improve it, and he couldn't do that in 1 or 2 years. It was indeed a very costly process."

At first the land-grant colleges were not much better than trade schools, teaching simple courses like "How to Plough." Then in 1887 Congress put them on a scientific footing when it passed the Hatch Act, which established agricultural research stations. These and the colleges worked together on soil science, plant science, diseases, entomology, agronomy, plant breeding, and animal husbandry. This produced a great accumulation of research. Even so, it was not until 1914 that the extension services were set up to get research findings out to farms.

Americans have always been better at pure scientific research than putting their findings to practical use as technology. This is also true of the British. Dr. Borlaug uses the example of Austrian botanist and Augustinian monk Gregor Mendel, who formulated the laws of heredity and laid the foundation for modern genetics through his breeding experience with peas. Dr. Borlaug: "You've got to make things happen. They don't happen by themselves. Mendel formulated the laws of heredity in 1865. But they never saw the light of day while he lived." I heard the same thing in London: Cambridge has more Nobel prizes than anybody in physics, chemistry, and physiology; discoveries like determining the structure of DNA have been practically applied better elsewhere. Yet in America in time the land-grant colleges did manage to bring results from the scientists' laboratories to the farmers' fields. This was a seminal process. It gave rise to global American supremacy in agriculture.

I've mentioned how culturally conservative farmers can be. When it comes to politics, generally they have clung to the belief that they can improve their own fortunes in the teeth of repeated economic depressions and a good deal of exploitation by banks, elevators, and railroads and, these days, big corporations with the clout to influence markets and the price of inputs. Historically, when the farmer did engage in radical or liberal or, more accurately, protest politics, it was a response from a specific ethnic group. As Professor Schultz points out, the La Follettes in Wisconsin spoke for small Norwegian farmers and for German socialists in Milwaukee; the Farm Labor Party of Minnesota was heavily Scandinavian, as was North Dakota's Non-Partisan League. The nearest thing to an agrarian national party was the Populist Party, formed in 1892, which gained momentum until the Democrats nominated William Jennings Bryan as their presidential candidate in 1896. Bryan's populist eloquence stole the thunder and the voters from the Populist Party and it faded away. Bryan's famous "Cross of Gold" speech at the 1896 Democratic Convention urged:

> Burn down your cities and leave your farms, and your cities will grow up again. But destroy your farms and the grass will grow in every city in the union. . . . You shall not press down upon the brow of labor this cross of thorns. You shall not crucify mankind upon a cross of gold. . . .

(Incredibly, this speech was recorded, surely one of the earliest times this was done. Studs Terkel dug it up and played it when he interviewed me on his Chicago radio show in 1986. It was an uncanny glimpse back into the nineteenth century.)

Another important reason the Populist Party failed was its inability to form a genuine rural-urban coalition. Agrarian protest had little appeal for wage earners of the industrial East. In the Middle West, industrialization only really came with the 1850–80 revolution in farm machinery. Wages were low, terrible slums were forming in fast-growing cities like Chicago, and newly rich and powerful manufacturers like Cyrus McCormick feared worker unrest. The *Communist Manifesto* was published in 1848. As the reputation of Karl Marx spread with *Das Kapital*—the first volume came out in 1867, the second and third volumes in 1885 and 1894—so did fear of his ideas. And of the Marxist war cry: "Workers of the world, unite! You have nothing to lose but your chains!"

In 1889 Henry Ford was just an ex-farm boy working as an apprentice in a Detroit machine shop. It was years before the apostle of mass production would found the Ford Motor Company and transform America by paying his workers $5 for an 8-hour day. The idea that America's urban working class could become a mass-consuming middle class was far in the future. So in 1889 McCormick, who gave $100,000, and other Chicago industrialists, instead of offering higher wages, decided to combat the Marxist threat by financing a campaign of Christian evangelism among their slum-dwelling, poorly paid workers. This was just 3 years after Chicago's Haymarket Square riot of 1886, in which seven police officers and four workers were killed. The violence and bomb explosion came after "anarchists" called a meeting to protest an earlier clash between police and strikers outside the McCormick Harvesting Machine Company, now turning out over a thousand new reapers a year. The strikers had demanded an 8-hour day.

Rather than give it to them, the industrialists turned to Dwight M. Moody, the great nineteenth-century evangelist, sort of the Billy Graham of his day. Moody was then at the height of his popularity, and he called for volunteers to come to Chicago in April 1889 to train for 6 weeks as "Christian workers" to save the heathen urban "masses" and quell any possible unrest. (This rather farfetched-sounding idea later appeared in French historian Elie Halevy's much debated "Halevy thesis," that the continuing political and social stability of the English was largely the result of nineteenth-century Methodist and evangelist movements.) The emphasis, Moody declared, would be on "the surrender of the will" among volunteers who would "sacrifice all for the Master." Nearly five hundred idealistic young men and women from all over the Middle West responded. One of them was my grandfather.

In his opening speech, Moody told his volunteers, "God sent you here. . . . The Haymarket Riot showed how urgent is the task. Either these people are to be evangelized or the leaven of Communism and infidelity will assume such enormous proportions that it will break out in a reign of terror such as this country has never known." The effect on Chicago is not known. The effect on my grandfather, Hadwen Williams, a 38-year-old doctor from Iowa, is that he came home, told my horrified grandmother he was giving up his medical practice and Quaker religion, and sat down and sewed an enormous tent for revival meetings. He spent the rest of his life as a Methodist evangelistic preacher.

The Worldwatch Institute's Lester Brown once asked me, since I had spent so much time in Third World villages, if my father had been a religious man. I had to laugh, since he was anything but: a hard-drinking, agnostic, half-Irish ex-farmer and country doctor who spent Sunday mornings on the golf course. Rev. Williams was not as intellectual or worldly but, in what you might call American Gothic style, was rather saintly in a rough-hewn way.

Fundamentalism sounds strange in our secular age and is tarnished by images of discredited TV evangelists less motivated by Christian impulses than by self-interest, sex, and greed. And mainstream Protestant churches have been losing members. But I think we can say most Americans are religious in one way that counts. They may or may not go to a church or a synagogue. But most people like to think their ethics are based on Judeo-Christian foundations.

My grandfather left behind notes for a great many of his sermons, as well as several speeches and a lifelong journal he kept with his wife; these shed light on how people tried to save farms and small towns early this century, particularly in the "country life movement" launched by Theodore Roosevelt. Although Roosevelt was born and raised in New York City, he became discouraged by a downturn in his political career and the deaths of his mother and his wife, and he spent 3 years, 1884–86, on a ranch in western North Dakota. The experience left him an ardent advocate of agrarianism, much in the tradition of Jefferson and Washington (and later, John F. Kennedy). In 1908 President Roosevelt set up a "commission on country life" to see how the flood of rural people into the cities could be stemmed. When Roosevelt reported the commission's findings to Congress a year later, he declared, "The men and women on farms stand for what is fundamentally best and most needed in American society."

The commission, which had held hearings all over the country, recommended better schools, better roads, parcel post, more credit, diversified farming, agricultural extension, rural cooperatives, and government control of monopolies—all things farmers had been demanding for years. It also singled out for special criticism country schools and churches, partly blaming them for "ineffective farming, lack of ideas, and drift to towns." Stung by this criticism, Grandfather roused himself to the country pastor's defense. At one country life conference at Ames, Iowa, in 1910, Rev. Williams spoke of the "steady exodus from the land" the previous 20 years. "Country churches are the despair of many a pastor," he said. "How often have I walked seven or eight miles on a Sunday evening just to face a few elderly souls whose young folks have gone to the city?" Some argued, he said, that rural churches and schools, together with stores and farmsteads, ought to be clustered at railroad stations, on the European model. He favored consolidating schools, but rejected the rest. "The course of American history suggests no such future peopling of the lands. It falls to the church to serve the farmer where he lives, and that is going to be in scattered homesteads."

He said the American moral code had grown out of rural life and reading the Bible, the Old and New Testaments providing rules for living. In a passage I particularly like, he said, "The Old Testament prophecies and sermons were preached by great souls made anxious by the exodus from the land into cities. Isaiah denounces practices common in Iowa today: the evils of absentee landlordism; the large numbers of tenants and sharecroppers; the idleness of retired farmers; the pitiful condition of hired men who must work twelve hours a day and seek refuge in the intemperance of the saloon. . . . Why, we've got landlords who won't tile poorly-drained land but who blame poor crops on tenants. I say to them, 'When you tile your land, you get your man.' Not before." Tile your land, sinners.

Grandfather was the son of a farmer and worked on a farm until he went to college (he had to provide his stern Quaker mother wages for a hired man until he was 21); this shows in his speeches. In one he argued that country preachers needed to go to "farmers' institutes" to keep up with all the new farming techniques: "No rural pastor ought to be ashamed if he can help a young farmer's wife dispense with parasites among her poultry or help her husband cure abortion among his cows. To know about soil conservation, balanced rations and improved livestock ought not to impair any man's ability to present a strong Gospel. . . ." One time he said he favored using a "road grader and King Drag" to improve muddy country roads. He was a great believer in education and observed, "Many an obscure pastor has earned a good record in the Book of Life helping farm boys and girls get into high school and colleges." He warned there was too much emphasis among Methodists on "large endowments, commodious new churches, and columns of converts," all of which made "the adding machine supreme." He saw scientific farming, good roads, and good schools as essential to rural revival. He did not realize these very advances would hasten the rural-to-urban shift.

Rev. Williams practiced what he preached. During the influenza epidemic of 1918–19, already an old man, he went back to his old profession of medicine, going about the countryside on foot, tending the sick and the dying. His health broke and he never fully recovered. He died in 1926 just a year after John Scopes was tried in Dayton, Tennessee, for teaching Darwinism.

I used to wonder why Grandfather left being a doctor to become a preacher. My father, whom nobody can remember ever going to church, said one time he guessed it was because Grandfather somehow sensed people in his day needed tent meetings because they didn't have diphtheria antitoxin. In the 1880s, my father said, every diphtheria epidemic kicked off a religious revival. A doctor knew how to relieve suffering, set bones, sew up cuts, deliver babies, and open boils. But not much else. Grandfather might have felt a preacher's sympathy was more needed than a doctor's skill.

By the late 1920s, the number of cars in America had quadrupled since World War I to twenty-eight million; two out of every three families had a car. When it came to transport, the age of horses was dying with the decade. It kept up much longer on farms, with horse-drawn machinery outlasting even a brief period of steam-powered machinery. Steam engines reached their peak in 1913, when over ten thousand were made. But steam-driven tractors never caught on very widely.

Steam is most remembered from the great threshing rigs. In North Dakota the wheat harvest began in August. It was a gigantic task. The fields gradually turned into a sea of shocks, curing in the sun. Threshing took armies of men. Dr. Borlaug recalls how in his hometown of Cresco in Iowa all the neighbors pitched in and threshed at each other's farms, common in those days.

Here's a description of steam threshing in North Dakota in 1912 from one of my uncles, now 95: "A neighbor always brought his threshing machine, a 38-inch separator, as well as a separator man, fireman, and engineer, all local people who went home at night. It took eight horse-drawn wagons to keep enough bundles coming. Another wagon hauled the threshed grain to the elevator. There were a few itinerant workers, but we mostly had local boys with their own teams and wagons. Threshing was a big undertaking. It could last until snowfall and needed strong arms and backs. But once you got going, it went fast. The bundle haulers worked steadily, going back and forth to the shocked fields in their wagons, and the pitchers had to hustle with their forks to keep the rig's chute full. Chaff blew into the air and the straw blasted out into dusty clouds and settled in a big light-colored stack. A stream of grain poured down to make a brownish-yellowish cone on the wooden floor of the grain hauler's wagon."

These big threshing rigs had vanished by the time I did farm work in the summertime, but my brother Jim, 14 years older, drove a horse-drawn bundle rack in 1932 when he was 15. It was a big rig from Minnesota with its own crew: a snoose-chewing separator man who was boss; an engineer; a man who ran the steam engine called Tankee; Straw Monkey, who carried away straw on his rack; twelve bundle haulers; four spike pitchers; several grain haulers; Jim said you might have thirty or forty men in all. "By 1937," he told me, "the last year I threshed, they were already using tractors and everybody was getting mechanized. It wasn't the same anymore." It was his last summer in Prairie.

In North Dakota's Red River Valley there was a good example of large-scale farming based on steam power. When the Northern Pacific Railroad suspended its construction of new lines in the 1873 depression, some company officials accepted huge tracts of land in Dakota Territory in exchange for railroad bonds. Huge wheat farms were established in the Red River Valley, some so big the plowed furrows were 6 miles long. Horses and mules supplied some of the power, but great big steam tractors were mainly used. What became known as the bonanza farms showed—as we see again and again—the problems that beset farming ventures that get too big: breakdowns of machinery, the unreliability of hired labor, the unpredictability of weather, low prices from too-abundant crops.

The same problems affect big corporations, whether state run or owned by stockholders, as affect small, family-run farms. But a family can reduce or defer spending. It does not have to meet a payroll or fill a government quota or show a profit to stockholders. In North Dakota by the 1890s almost all of its gigantic, single-crop farms had given way to small, diversified holdings.

I've seen the same problems defeat huge commercial farms in Khuzestan in southwestern Iran in the 1970s—the Shah wanted to "make the desert bloom again" as in biblical times; in Sudan's Nuba Mountains—a World Bank–funded fiasco as local Arab moneylenders, not the Nuba farmers, had the capital to invest; and in parts of China.

Collective farming is utterly discredited in Russia too. I think we can look ahead to huge increases in food production now that Russian farmers are getting permanent guaranteed tenancy to land they till. One has to be slightly cautious because Russia's farming failures can also be blamed on its very northern latitudes and variable climate, early frosts some years, and a tendency to low rainfall and little irrigation. But Stalinist methods, like big corporation-type farms, inhibit scientific advance and do not provide enough incentives to cultivators. As Aristotle said long ago, communism breaks down because of man's natural inequality; it provides no adequate incentive for those of superior ability. Aristotle believed the stimulus of gain is needed for hard work, just as the stimulus of private ownership is needed for industry, husbandry, and care. He wrote: "That which is common to the greatest number has the least attention bestowed upon it. Everyone thinks chiefly of his own, hardly ever of the public interest."

Look what happened in China when Deng Xiaoping brought back the family farm (and expect China, once its elderly rulers leave the scene, to make another headlong dash for freedom). In 1979 Deng introduced cash incentives for surplus grain and cash crops. In 1981 he broke up the communes and restored the family farm as the basic production unit. In 1981–84 China's grain crop shot up nearly 100 million tons to 407 million tons, the all-time world record for a single nation in a year. Average rural incomes rose 70 percent. A combination of wise politics and science made it happen. Machines mattered little. Most of the record 1984 harvest was grown the age-old way, with hoes and sickles, sometimes without even a plow, 60 percent of it by women. Seeds, water, and fertilizer, plus incentives, were what mattered. The seeds were Chinese versions of Dr. Borlaug's high-yield, dwarf, fertilizer-intensive, fast-maturing wheat and rice, such as have transformed tropical farming since the mid-1960s. Water was important, and here Mao Ze-dong gets credit; under Mao China's irrigated land rose from 20 percent to 50 percent of total farmland—which is about three times what we have.

The big energy source of American farms is, of course, the oil-fueled internal combustion engine. The first gasoline-powered tractor was built in 1892. As with horse-drawn machinery and the Civil War, it took World War I before tractors widely caught on. It was the same with combines. A horse-powered combine, which both cut grain and threshed it in a single operation, was built in Michigan in 1836 but never was used widely. Large combines powered by gasoline engines did not become widely available until the 1920s. It was not until the late 1940s that the modern combine generally replaced both the stationary threshing machine and the grain binder, which had mowed the wheat and bundled it into shocks. Now the gasoline-powered combine, far more efficiently than anything that had come before, could cut the wheat, gather it up, thresh it, separate out the chaff, return the straw to the ground, and collect the grain in a hopper for delivery to a truck, all with one machine. Naturally the combine at last was used by everybody.

Such mechanization makes little sense, of course, in the Third World with its cheap labor, costly oil imports, and so many mouths to feed. In India in the stables of a maharajah, I once came across a rusted old tractor of the 1920s, a long-forgotten status symbol that was never really needed.

What has happened in farming, of course, reflects technological change on all fronts. Take medicine. Every year since the 1930s has made it more of an exact science. This leaves little place for the old-fashioned doctor with his intuitive ability to diagnose, born of hard-won experience. The specialist, the laboratory, all the impersonal scientific techniques of a modern hospital have taken over. Both my grandfathers knew how to set bones, sew up cuts, and deliver babies, and not much else. Only a few therapeutic drugs existed during the first third of this century. But what a doctor lacked in scientific knowledge was partly made up in practical experience and warm relationships with patients. As in farming, every advance has its tradeoff. Dr. Borlaug says, "The old-time country doctor knew how to stimulate the patient's own innate defenses. If a person got too depressed, his whole immune system weakened." With sulfa to fight infectious diseases, soon to be followed by antibiotics, medicine made its turning point 50 years ago. (And world population exploded.)

In education in America, it was not until the Bennett Law of 1879 that every child was required to spend at least 12 weeks in school each year. Four subjects had to be taken: English, reading, mathematics, and history. Emphasis was given memorization of historical dates, multiplication tables, combinations of numbers, verses from Shakespeare and the Bible, and penmanship. These tales, whether biblical, classical mythology, fairy tales, or patriotic legends, gave children something to live up to. Memorizing poems, particularly, is important. What makes a community live is its shared remembrance of culture and language. It was good education. Here are some questions from the 1905 examination my mother had to pass at age 17 to teach in a one-room country school in Iowa: Put in the punctuation for Hamlet's "To be or not to be" speech; name five important events in Andrew Jackson's administration; name the fluids secreted by the digestive system; locate Sitka, Veracruz, Glasgow, Tokyo, Madras, Santiago, and Timbuctu; define "molecular force"; and find by division the G.C.D. of $15a^4 + 10a^3 + 4a^2b^2 + bab^3 + 6a^3 + 19a^2b + 8ab^2 - 5b^3$. Just about the only one I got right was Timbuctu, and that was because I'd been to Mali. A weakness of present-day American and British education, I've found in visits to schools, is the emphasis in learning access to knowledge through computer skills, but not memorized knowledge. Who learns poetry by heart? Both are needed. The speech of older Americans can be rich in unconsciously quoted phrases from Shakespeare and the Bible, allusions almost totally gone from the speech of post-television younger people.

Early this century my grandfather, the preacher, found traveling performances of *Ten Nights in a Barroom* and *Uncle Tom's Cabin* acceptable for his Methodist flock because they provided "moral lessons." The managers of such touring companies shrewdly took pains to advertise: "A High Tone of Morality Will Be Rigidly Maintained." Uncle Tom might be flogged by Simon Legree, but he would go to Heaven. Eliza might be pursued across the floating ice by savage bloodhounds, but she would always safely reach the Ohio side and freedom. Movies, which rarely were tailored for rural audiences, or concerned to provide their children with rules for living, bothered Grandfather from the start. (Grandma, who adored them, used to sneak into them incognito, hoping no one would recognize the Methodist preacher's wife.) Grandfather, after seeing a few, wrote a friend in 1910, "One cowboy reel packed more violence than any dime novel would dare to print. I've come around to thinking that those folks in New York who make the films must have one idea only: how to provide thrills. Hate thrills, love thrills, nothing but thrills. Intemperance, gambling, wild chases, firing guns right at the audience, they engage in every sin a person can think of." The movie industry soon moved to Hollywood. But 80 years after *The Great Train Robbery* it is still, figuratively, firing guns at the audience.

The problem of our mass-media fantasy culture, then and now, it seems to me, is that so much of it is dreamed up by highly talented people in Hollywood or New York whose main interest is to entertain and make money. Of the twenty-five leading fiction best sellers of the 1980s, six were written by Stephen King. Even Barbara Cartland, the world's all-time best-selling author, complains about our "permissive era" when people write like her, but with pornography. "You can see where they've slotted it in," she told me over tea in London. A Dickens or a Mark Twain also aimed to entertain, but like Miss Cartland (her heroines always get married before they go to bed) they felt they had to inculcate moral lessons too, to provide their readers with practical rules for getting along in real life. One can name exceptions from *All Creatures Great and Small* to family sitcoms like the Bill Cosby show, but in the main, the media in America fails in culture's basic role: to provide a workable design for living. It fails to give young people a reassuring background against which their own lives take on meaning.

The faster change comes, the less the old solutions and rules apply and the more new solutions and rules have to be invented, which can be a worrying and confusing experience. This century, change has steadily accelerated. When my father farmed as a young man, working 320 acres, or a half section, in 1910–17, the average American farm size was 148 acres. In arid North Dakota, farms had to be about twice that size; my father grew wheat, oats, barley, rye, and corn. He grazed cows in a 60-acre pasture. Pigs rooted and squealed in a lot, and chickens clucked underfoot, running free about the farmyard. Down near Cresco, Iowa, on much richer land, Dr. Borlaug's father managed with just 56 acres, going up to 105 acres just before the 1929 Crash.

The grandson of a Norwegian immigrant, Norman Borlaug, born in 1914, grew up in an all-Lutheran, Norwegian community, essentially just a crossroads with some stores, Saude. My grandfather preached at one time or another in the nearby towns of New Hampton, Lime Springs, Monona, Postville, and Ridgeway. My mother taught in a one-room school in Hardin, 40 miles from the Borlaug farm, and for 3 years in Elma, just 10 miles west. Dr. Borlaug recalls:

"The little one-room country schoolhouse I went to up there—of course that was in the twenties—it was still half-Norwegian, half-Bohemian—they didn't like to be called Czechs. Just to the north, the second little town over, was Spillville. That was pure Bohemian. Antonín Dvořák spent a summer there to capture the feeling for his *New World Symphony*. 'Going Home,' that's one of my favorites. Our farm, Dad owned just 56 acres and the rest was Granddad's that he later rented, another 49 acres. We had chickens. Sold pigs and eggs for cash. There was a little crossroads, Saude, with a co-op creamery, general store, blacksmith and machine shop, feed mill for grinding your oats and corn, the Lutheran church, and that was it. You'd take eggs and buy what you needed. If they were worth more than what you presented, you got a 'due' bill. It showed how much you got for the eggs, how much you bought, and what you had due. In the winter you'd go down to the river and haul back a sled-load of ice. Each fall we'd sell six or seven head of beef, drive those cattle 13 miles into Cresco. Generally you and the neighbors at the same time. All Norwegians. With the Bohemians on one side and the Irish on the other."

All over the Middle West the first half of this century, on such farms, the men got up about five and headed for the barn to clean out the stalls, feed hay and oats to the horses, water them, and feed and milk the cows. My mother, once she went to North Dakota and married my father, who was then farming, didn't milk, but like most women, she separated the cream; like butter and eggs, it was the big source of cash. Butter was usually churned once a week. Skim milk went to the calves or hogs, who also got a few bushels of ear corn.

While the men curried and harnessed the horses, women fed and watered the chickens. You needed a hearty breakfast after a couple hours of hard work, and there were usually plenty of eggs, ham or bacon, fried potatoes, homemade bread, jam, and coffee. Dr. Borlaug remembers johnnycake, potato cake, and wheat cake. After breakfast my father and his younger brother used to hitch four horses to one plow and three to another and head for the fields. They felt that 20 miles with a gang plow was a big day; it meant they'd plowed 5 acres.

At noon horses were unhitched, watered, and fed before the family ate. The same at night. Chores—milking; separating; feeding horses, cows, hogs, and chickens—went on again until supper around seven. By nine thirty, after a final trip to the outhouse, everybody was ready for bed. So it went, day in and day out, 7 days a week, the routine varying if there was planting, haying, harvesting, or threshing. Cows had to be milked morning and night 365 days a year. Women killed and plucked their own chickens, baked their own bread, made clothes on a treadle sewing machine, and canned much of the family's food with a pressure cooker and 2-quart mason jars, including chicken, pork, beef, beans, peas, corn, tomatoes, and potatoes. "Not much fruit," Dr. Borlaug recalls. "We used to get an orange and a little sack of candy when we sang carols at the Christmas tree in the Lutheran church." (This is an exquisite little church. It still stands out in the country in a grove of pine trees in what looks like a landscape in Norway.)

The first car in Dr. Borlaug's family was a Model T in 1914. My father didn't buy his first car—also a Model T—until he began to practice medicine in 1922. Rev. Williams owned just two cars in his life, a secondhand 1904 Ford he bought in 1910 and a brand new 1915 Model T he drove for 11 years. My father's first tractor was a Fordson four-cylinder, 20-horsepower machine he bought in 1917. He used it just for plowing, as his drill and binder were set for horses. Dr. Borlaug: "The first functional tractor we had was a little Fordson in 1926 or 1927. Oh, we still had horses, kept them right through World War II. I had a bachelor uncle, Oscar; he was a damn good mechanic. And he bought some of those old very first tractors. No tires. Just on steel. They never were any good. He bought a Waterloo Boy and an Avery. They'd been junked someplace. And he had 'em out in the woods. For enjoyment, he'd be out there on Sunday afternoons, trying to best those things into shape and they'd be completely hopeless."

The implement dealer who sold my father his first tractor, a man now in his 90s, told me North Dakota's farmers didn't start to buy large numbers of tractors until after World War I. A lot of the earliest ones were Caterpillar and steel-wheeled, some of them steam powered. Everybody kept on horses. By October 1929, there were still only 378,000 tractors in the whole country, though by then they had rubber tires. Just 50,000 tractors more were bought during the first 5 years of the Depression, which also saw plagues of dust storms and grasshoppers. It was not until 1935–40 that the total suddenly rose to just under one million. During the war everybody shifted from horses to tractors. So mechanized farming in America as we know it is just about 50 years old.

As mentioned, widespread application of biology and chemistry to farming began about the same time. Nitrogen fertilizer production tripled in 1920–30. Hybrid corn swept the Midwest after 1933. Rust-resistant wheat, improved cotton, better hog and dairy cattle breeding, and higher-yield forage crops all date from the late 1920s. With the New Deal came such federal programs as the Agricultural Adjustment Act of 1933. From then on, the government was deeply involved in farming. The reason so many of its programs aimed to help small and medium farmers but ended up benefiting bigger commercial farmers more is that crop supports were favored over income supports.

It all meant vastly increased farm production. In 1940–80 the combined output of seventeen major food crops rose 142 percent, from 252 million to 610 million tons. Corn yields alone rose 251 percent. Dr. Borlaug attributes the spectacular rise in American farm output to new high-yielding crops, more irrigation, a fourteen-fold increase in fertilizer use since 1938, better weed and insect control, and sufficient economic incentives so that farmers keep adopting new methods. He credits the agricultural universities with coming up with the basic science but says private businesses did more to get it applied as technology in the farmers' fields.

Such heavy reliance on chemicals and machinery has its price, of course. Since World War II alone, there has been a tenfold increase in herbi-, pesti-, fungi- and all the other cides, some of them toxic, used on the land. Some seeps into groundwater. In southern Iowa, for instance, many farmers have stopped drinking from their private wells. They have banded together to sink communal wells and to install the expensive equipment needed to remove chemical contaminants. Yet farmers oppose limits on the use of chemicals unless they get compensated for loss of income and are given incentives to encourage voluntary compliance. The Reagan administration's stated ambition was to eliminate all farm subsidies by 2000. Some farm lobbyists, in effect, want to convert environmental initiatives into still more forms of subsidy.

Today's intensive farm methods can also compact the soil, reduce organic matter, and endanger wildlife when plowing. When land is farmed, in the words of former Secretary of Agriculture Earl Butz, "from fence row to fence row," the price is paid, not just in adverse environmental impact, but because farms get ever bigger and fewer. It is hard to even get into farming anymore. Don Paarlberg, professor emeritus of agriculture at Purdue, says the traditional "farm ladder" has been snatched away. He says, "My father started farming 80 years ago with a team of horses, a few tillage tools, and about $200 in savings. He retired 40 years later, the unencumbered owner of a moderate-sized Indiana farm. He climbed the agricultural ladder: first farm laborer, then renter, then part owner, then full owner free of debt. He was hardly unique; all of his six brothers started farming in similar circumstances, and all became debt-free operators. My father expected each of his four sons to follow in the same path, but they did not. Why? What happened? Were the sons less able than the father?"

Paarlberg's explanation lies in the way farming has changed. Today just about the only farmers who can afford to buy land are those who own land already. The only way to acquire a farm is to be born the child of a farmer or to marry a farmer's son or daughter. To enter farming from scratch today without family help might take $100,000 of equity capital, even if land is rented, money borrowed, and machinery bought secondhand. (Yet we will see two striking examples of just that.) Populist economists like Paarlberg, who for years was the Department of Agriculture's chief economist, fret about a landed gentry taking the place of American family farmers. They warn we might have to accept a wealthy hereditary farm-owning class. More radical economists want root-and-branch reform of tax laws for agriculture, arguing they are invariably biased in favor of big landowners. I've mentioned how, among farmers themselves, hardly anybody wants to put limits on individual holdings. Consolidation will have to go a lot further before this deeply ingrained American attitude changes. In the meantime, as we urge land reform on the rest of the world, we may need it worse at home. But imposed land ceilings are anathema to most American farmers.

It used to be that on a typical family farm, the farmer, his wife, and their children supplied most of the land, labor, capital, and management, with a little borrowing and renting along the way. Farming on these terms was a possible prospect for a farm-raised youth. No more. As we will see in Prairie and Crow Creek, the two rural communities we are about to turn to, land, labor, capital, and management for a medium-sized farm have to be split up in some fashion. We will have examples of each of the ways this can be done.

A farmer can borrow. In the 1970s this seemed like the answer: borrow, leverage yourself, and so acquire a big enough farm to be efficient. Low interest rates and rising land values made it seem like the solution. But the 1980s showed heavy borrowing can expose a farmer—and the bank that lends the money—to unacceptable risk. Over time the return on farmland can be less than the mortgage rate of interest. The desire to own land pushes prices up to levels with low rates of return.

Off-farm income is another way. Since 1945 the proportion of income to farms from off-farm sources has risen from 26 percent to more than 50 percent of net income. By 1987, just over half of all employed people living on farms worked principally outside, reaching nearly 75 percent for women. Among farm men, the increase since 1950 was from 26 to 38 percent, a real but not radical change. An equally important trend is the rapid rise of farm women in the labor force, up from 23 percent in 1950 to 55 percent by 1987. With so much off-farm work, some tension exists. Full-time farmers can argue that part-timers subsidize their farming in unfair competition. A farmer who earns $10,000 a year in an off-farm job doesn't need as much land and can custom-hire his machinery. Many full-time farmers look on off-farm jobs as an admission of failure. But as can be seen in communities like Prairie and Crow Creek, more and more of them are making the change.

Renting is the most common way of entering farming for those without capital; as I mentioned, Prairie and Crow Creek both have striking examples. Only about a third of American farms are now operated by full owners (many of whom have mortgages). About 10 percent is farmed by tenants. Fully half is farmed by part owners who, typically, own half the land they farm and rent the other half. Renting, particularly share-renting—either fifty-fifty or one-third–two-thirds—cuts down the tenant's risk from price fluctuations in land and crops. But in tough times, as in the 1988 and 1989 drought years, many landlords insist on cash rents. And most renting contracts are from year to year. How does a tenant make long-term plans? As we'll see, it is hard. Laws for long-term leases or to protect the value of the renter's improvements have yet to be made.

A fourth option is to farm on contract. Farmers may custom-feed somebody else's cattle. Or grow hogs with equipment supplied by the packer. Or family farms can be incorporated, to avoid the problem of one heir having to buy out the others; in time this shifts ownership from farm to nonfarm people.

If a farmer fails to split up land, labor, capital, and management successfully, he eventually goes broke. The number of farm people is dropping fast, from 6.1 million to just under 5 million in 1980–87 alone. Losses in the 1950s and 1960s were even higher but aroused less public concern, mainly because they mostly were small farmers, especially tenants. Since 1950 practically all black sharecroppers have been forced off the land. At the same time, white tenants, just over 1 million of them, or 70 percent of the total, left farming too.

In contrast, so many of the farmers who are failing now are well-educated, technically efficient younger operators. And small towns were harder hit in the 1980s than ever before.

All the losses have meant a big shift in America's farm population from the South (52 percent in 1950, 29 percent now) to the Middle West (32 percent in 1950, 51 percent now). Until 1970 farmers were aging, with those 60 and over twice as many as those 35 and under. Today their numbers are about equal.

The pattern of ownership has changed drastically. Much of the Middle

West was given to family farmers with 160-acre tracts under the Homestead Act of 1862. This meant widespread farm ownership and farms fairly uniform in size, usually four to a section and marked by dirt roads (a pattern you still see when flying over). Today in the north central states, 44 percent of farmland owners own less than 50 acres and a total of just 7 percent of the land. A mere 0.3 percent own over 1,000 acres each and almost a quarter of total land. This would touch off a revolution in a Third World country where 75 to 80 percent of the people are peasants who grow much of their own food. But just 2 percent of Americans now farm. Who owns what and how it gets done matters a lot less to the other 98 percent than having a cheap, reliable food supply at their local supermarket.

So the populist rural economy on which so much of American culture is still based has in reality largely come and gone. It peaked in the 1930s just before World War II; the mechanical-chemical revolution has meant a steady, only semivoluntary, exodus from the land almost ever since. Calvin Beale reckons in 1949–60, 21.5 million Americans left farming, an average of over 1 million a year. The farming population dropped from over 30 million to below 10 million. In 1950 there were still 5.6 million farms with an average size of 214 acres; in 1990 there were 2.2 million with an average size of over 400 acres.

The small holdings with sales of $1,000 to $20,000 a year make up 70 percent of these officially classified "farms," but they account for just 10 percent of total farm sales and their operators earn 85 percent of their incomes away from farms. The 6 percent of all farms—162,000—that have sales of over $100,000 a year make up over half of total farm earnings.

Until the boom ended in 1981, bigger farms were absorbing smaller farms at an ever-faster rate. For many farmers in communities like Prairie and Crow Creek, the mechanical-chemical revolution in the 1970s became economic Darwinism at its most gory. Those who did not go into debt to buy gigantic new machines, did not sow the newest hybrid seeds, did not learn to adapt their marketing to bouncing prices on the Chicago exchanges, and did not buy or rent neighboring land to reap the economics of scale either had to get out of farming altogether, isolate their farms from the community's economy, or take part-time jobs.

We'll see examples of all three. In Prairie we will talk to sixteen farmers, including three successful ones with sidelines (beef cattle, a seed plant, and a turkey farm), two badly in debt, a big renter, a big landowner, a small and old-fashioned farmer, two part-time farmers, and six or seven farmers' sons. Among nine retired people interviewed, six of them are ex-farmers or widows. We will also meet eleven actively working nonfarm people in Prairie: newspaper editor, health worker, auctioneer, bureaucrat, sheriff, banker, Catholic priest, Lutheran pastor, two Sioux Indians, and Hispanic migrant field worker.

In Crow Creek, which is only half Prairie's size and within commuting distance to a city, we will hear from only four farmers, but again a big operator in debt, a big renter, a small and old-fashioned farmer, and a part-time farmer, all types met in Prairie. Among nine retired people interviewed, two are farmers' widows, but this time two are ex-factory workers. One farmer's wife raises poultry, but another commutes to a delicatessen job in the city; we will also meet three commuters to an urban avionics plant and a Methodist pastor, himself a city man. Everybody has been affected by the crisis in farming.

By the 1970s, to plow, disk, harrow, and cultivate had come to mean sitting in a giant tractor with a crush-proof cab, power steering, air-conditioning, and stereophonic radios, plus sensing devices to tell the farmer if the flow of seed, fertilizer, herbicide, fungicide, or insecticide was interrupted. When the crop was finally in, some farmers, no longer tending livestock, took off for a few winter months, going to Arizona, Texas, or Florida. Those with livestock were still tied to the land; animals have to be fed. But many just grew grain crops; they had a lot more leisure.

The 1970s, too, saw cultural attitudes subtly change. When I was growing up, debt was sinful; now it came to be seen as essential for tax purposes and to cope with inflation. Farming was skillfully putting other people's money to work. Farmers became avid market watchers. In Prairie and Crow Creek I was to find quite a few who read *The Wall Street Journal* for its daily reports on the commodity market or subscribed to electronic news services on minute-by-minute farm price movements. Everybody tuned into radio reports. In, say, 1980, one might have forecast the future of the American farm as a big commercial enterprise, still run by a family, still relying on the oil-fed combustion engine and oil-based fertilizer and chemicals, but with more computerization, automation and robotics, and city-style comforts.

Now, in the 1990s, the financial stress of even big farmers, rural bank failures, the need to cut the deficit and big federal spending, looming environmental issues, 2 years of drought, and rising world competition have put this future in doubt.

Even so, it is well to remind ourselves that farming remains America's biggest business. Its workforce is still as large as that of the steel, car, and transport industries put together. It is also America's most successful business: farm labor productivity has risen tenfold the last 50 years. Foods and fiber are still our biggest exports. Agriculture is the basis of everything else.

There will be good times ahead. Yet any return to long-term rural prosperity is ill-founded. The fundamental farm problem remains as unsolved as it has been for most of the twentieth century. In most years American farmers are able to grow far more wheat and corn than other rich countries are willing to buy or than poor countries can afford.

With low returns on investment and laws against big corporations coming in, the family farm, with all its worries about the weather and crop prices and damage to the environment, will survive. But how many little towns will? Can anything save them?

The best hope, their people feel, for a Prairie that solely depends on agriculture to avoid extinction, or a Crow Creek, now half commuters, to avoid getting culturally swallowed up by a city, lies with science and technology. Its machinery and chemistry and plant breeding got us into this fix; why can't its biology get us out? Where will the breakthroughs come? Hybrid wheat, cotton? Reduced tillage, biological controls, sterile insects, antitranspirants, drought-resistant crops, enhancement of photosynthetic efficiency?

If science—biology, physiology, biochemistry, immunology, genetics, and microbiology—is tied closer to agricultural research, it may be possible to maintain yields at lower costs and with less environmental impact. Science itself might provide a high-technology economic basis in new crop use and new land use to enable Americans to live comfortably and productively and preserve their rural culture in places like Prairie and Crow Creek.

Our urban culture depends on it.

But such hopes lie in the distant future. What is happening in such communities now?

# II

# *Prairie*

T HE NORTH DAKOTA prairie, when you see it for the first time, looks flatter and emptier than any landscape you have seen before. My mother used to say it was as if God had leveled it out with a rolling pin. Railroad tracks stretch into infinity. The sky is bigger and bluer, with row upon row of white cumulus clouds going on and on for improbable distances. The lone hawk wheels overhead, great flocks of seagulls turn a field white, Canadian ducks and geese fly north in long, wavering formations. There is nothing quite like it, though I was struck by a sense of recognition the first time I saw India's flat and featureless Punjab Plain. Chekhov's descriptions in *The Steppe* sound familiar too—"You can go on and on, and never see where this horizon begins or where it ends."

The little town I'll call Prairie, where my family happened to be living when I was born, sits right out in the middle of it. That was at the height of the Depression, 1931, and we moved to Fargo after 15 months. But my family never really left Prairie behind, so many of our best lifelong friends came from there. And before we went, my mother used to wrap me in blankets and a buffalo robe and leave me for hours in my buggy on our front porch, even in the dead of winter. Everybody said I'd freeze, but she said I liked it. So for me, I guess, the North Dakota prairie is bred deep in the bone.

Going back in middle age and driving about the countryside was to see it wasn't quite as flat as I'd remembered it, particularly the Red River Valley around Fargo where I used to bicycle along dusty roads through wheat fields full of grasshoppers. Around Prairie you can see the *coteaus,* or low chains of hills, rising in the southeast out of the broad limitless plain. Cross the James and the Sheyenne rivers going south, and ahead in the faint blue

47

distance is the Hawksnest, a high flat-topped hill the Indians call Huyawayapaahdi, or, roughly, the place where the eagle takes home its prey in its beak.

Hawk County, where Prairie sits today, was a hunting grounds for the Sioux until almost this century. John Running Horse, an old man of about 80, with a wrinkled, seamed, darkish face but still broad shouldered and healthy, who wears a slouchy old Indian hat and his coarse black hair in braids, told me what it must have been like. "In the early 1900s or late 1800s," he said, "the federal government decided to make farmers out of the Indians. So they set them up with a team of horses, a wagon, one of them one-horse cultivators, and a walkin' plow and a drag. Now when they say the Indians failed at this, they had a reason to fail. And the federal government here in North Dakota went out to round up all them wild ponies, wild horses that never had a harness on, and come up here and made a deal. And they brought wagons and hooked those wild broncos to the wagon. Before they could get on the wagon, the team probably kicked their heads off."

Running Horse, who speaks in a calm and deliberate bass voice, says the Indian had no tradition of owning land. "The Indian, he don't understand what a deed is. He's got this paper. He can't read or write. And the white man could convince him. So by thumb print, he's made a deal. Whatever that deal may be. What he realized from this thumb print could be a quart of whiskey. And 80 acres of land went. The Indian didn't know this. He cares less. 'Cause he's not a person that was acquainted with owning. They didn't have no right equipment. Nothing to seed with and no instructions. The Indian's no farmer to begin with."

Running Horse remembers, as a boy of 10 in his Indian village—that would have been in 1910—how white farmers kept coming in and buying up land. "That was something we could see," he says. "We knew it was happening when we'd see a wagon coming into the village, that a home was gone. Word got spread that Indian land can be gotten for a song. By hook or by crook, a farmer can get that land. Or the banker can buy it. He'd say, 'You sell me the land and I'll put your money in my bank and you can draw so much a month.' And when the Indian would go back to get more money, the banker only would say, 'Your money's all gone.'

"Before this was free land for the Indian people to live on. Now they got caught up in this. Now we're learning fast." Many Indians have moved into the mainstream of American life. He himself, he says, lived in Chicago for 5 years, working in a soybean plant. "That was kinda 5 years of hell for me. The job was easy. The Indians are in the big city. They're there, but they're dreaming reservation. Wherever they're at. The Indian, he dreams of wildlife, timber, deer, birds, ducks, lakes, rivers, wild rice, fishing, the whole life. Send him down to Minneapolis or Chicago and he's lost."

The Sioux held most of the land in Hawk County; Chippewa ranged only as far south as the Sheyenne River near the county's northern boundary. Government surveys began staking out sections and quarter sections in 1882, just 6 years after Sioux and Cheyenne warriors defeated Custer's forces in the Little Bighorn Valley. Sitting Bull, who led the Sioux, escaped to Canada, returned on promise of a pardon, and was settled on a reservation. But after a stint with Buffalo Bill's Wild West Show in 1885— my Iowa grandparents saw it in Chicago—Sitting Bull continued to champion the Indian cause, encouraging the Sioux to refuse to sell their lands, and keep to the old ghost dance religion. He was killed in North Dakota by Indian police on a charge of resisting arrest. The last major battle in the Indian Wars came at Wounded Knee in South Dakota in 1890 when, after a Sioux warrior pulled a gun and wounded an officer, U.S. cavalry troops opened fire, shooting nearly two hundred men, women, and children.

After Wounded Knee the ghost dance religion of many Sioux and Plains Indians was outlawed until 1978, an astonishing fact I only learned from a young Sioux, Stan Morrin, in 1990 in Prairie. Outlaw a religion in twentieth century America? Morrin is a handsome young man of 28 with a thin darkish-pale face, a high beaky nose, heavy eyebrows, and the blackest kind of eyes; his straight black hair hangs to his shoulders. Morrin trained to be a Catholic priest, then gave it up. "They slaughtered all those people at Wounded Knee," he told me, "because they were afraid their religion would unite them. You see, all the races of man are given a religion, a belief so the world around them and their lives have meaning. And they have a way to exercise that belief through prayer and worship. We were given that too. You love your neighbor. You love the great spirit.

"But the Indian carries it further. He says: Respect that rock. Respect the water. Respect the sky. Respect the earth, which is your mother, which all life comes from, where we go back to. If you outlaw Indian religion, as they did for so long, you get parents who aren't able to understand the old ways and aren't able to pass them onto their children. That takes us to the whole history of boarding schools, mission schools, ripping off Indian land, making Indian people something they aren't, destroying their culture, destroying their drums, destroying their whole being."

Attacks on a culture are always risky. I can remember how one time in a village in Sudan's Nuba Mountains the local school teachers had the children march around the town chanting, "Science not superstition!" The old *kudjur,* or rainmaker–faith healer of the tribe I lived with, when he spoke of "schools" and "modernization," like young Morrin, made them sound like curses. The *kudjur* once told me, "Now the harvests are poor and the land is not fertile and the rains are few. Why? The people have become Muslims. They are leaving the faith. They do not follow the god of our ancestors." Interestingly, even the local Muslim missionary among the Nubas, a Cairo-educated man named Sheikh Idris, felt cultural change must come slowly if it is up against a deeply ingrained tribal religion, which is also a way of life. "The coming generation, the children now entering school," he told me, "will not hold to these old traditions. But at the same time I cannot predict the outcome will be in favor of Islam. The real danger is that these people will become altogether godless. The Nubas could lose their old culture and belief in any god and have nothing at all." As I did, Sheikh Idris found African tribal religion extremely moral. "No one escapes a system of punishments and rewards from the moment he is born until he dies. For to do wrong in the tribal religion is to die before one's time or suffer some other evil fate."

The Indian past is hard to imagine around Prairie, where all but marshes, ponds, and river banks is now planted in wheat, barley, sunflower, bean, or other crops. Farther west in North Dakota, in the rugged rangeland Theodore Roosevelt said looked like "hell with the fires turned out," you feel you might expect to see buffalo and the spirits of Indian braves. But with so many windbreaks of old, unwatered trees where homes and barns once stood, Hawk County around Prairie is haunted by other ghosts. The little towns are the worst. So many abandoned buildings, so weathered and gray the grain in the siding stands out in relief. A door hangs on a rusted hinge and creaks in the wind. A tumbleweed blows by. The sky looks immense and limitless, as it must have looked to the Sioux, its great white cumulus clouds drifting across a far horizon.

"You mean all these huddled derelict buildings used to be little towns?" I asked a local farmer who was showing me around.

"Yessir. You see that spire over there? Well, that used to be a Lutheran church. Oh, it's nothing but an empty shell now, ought've been torn down long ago. But in the old days that church was the foundation and cornerstone of our family's whole way of life."

It was a way of life that came and now is going fast. Fifty years to grow. Fifty years to decline. As Louise Erdrich writes in the opening passage in

*The Beet Queen,* what mattered in North Dakota's settlement was the railroad: "Along the track, which crossed the Dakota-Minnesota border and stretched on to Minneapolis, everything that made the town arrived. All that diminished the town departed by that route too."

Half the land in the southern half of Hawk County went to the Northern Pacific in its land grant. The Northern Pacific built its line so far in 1883, and where the work stopped, the little settlement of Sykeston sprang up, the first town in the county. In 1892–93 the Soo Line built its main tracks into Hawk County and set up the towns of Cathay, Emrick, Fessenden, Prairie, Manfred, and Harvey. The idea was that no farmer ought to have to drive his team of horses more than 3 miles to reach a market town. Each settlement wanted to be county seat. A ballot was taken, Prairie won, and its men had to raid Sykeston with wagons, teams, and shotguns after dark to carry home all the land records.

The Northern Pacific and Soo railroads advertised that quarter sections of fertile Dakota Territory land were available. They mainly attracted Norwegian and German-Russian immigrants. In the 1890s the Northern Pacific's branch line was extended, and the towns of Bowdon, Chasely, Hurdsfield, and Heaton came into being. It was not until 1912 that the Great Northern railroad also built a line across Hawk County and added a third and final string of little towns, this time all with German names: Bremen, Hamberg, Heimdal, and Wellsburg. Some of these railroad-established towns grew; Harvey, with the county's only hospital, is more than 2,500 people today. Many of the towns never amounted to much; in 1990 all are in decline and a few face extinction. None has completely died away, but Emrick was down to four people in 1990.

In its century of settlement and desettlement, Hawk County went from Sioux hunting grounds and no farms in 1882, up to 1,670 farms in 1935, to just below 700 today. These cover 630,000 acres of cropland, worth as much as $1,000 an acre in the late 1970s, down to an average low of $341 an acre in 1985. In North Dakota itself, the total number of farms peaked in 1933 at 86,000 with an average size of 500 acres. In 1990 there are about 32,000 farms averaging just a little over 1,200 acres.

There are just over 660,000 people in the whole state, about 10 percent of them in Fargo. North Dakota lost 15,000 people in 1984–87, an exodus that has slowed. The population of the state's thirty-nine most rural counties keeps dropping; in its fourteen most urban counties it keeps rising. With America's highest death rates and lowest birth rates, North Dakota has just about the same number of people it had 40 years ago, though its number of Indians grew by 50 percent to over 20,000 in the

1970s, reflecting more decent living conditions on the reservations, even if joblessness is still severe. Most of the Native Americans are concentrated in a few areas of North Dakota, like the Devil's Lake–Fort Totten region just north of Prairie, and Turtle Mountain farther north or Fort Yates in the south. Many live outside the reservations to get work. With 9 persons per square mile, North Dakota, after Alaska and Wyoming, is one of the emptiest stretches in America.

Forty years ago North Dakota produced just a few crops: Hard Red wheat, durum wheat, rye, barley, oats, flax, and potatoes. Today these are still grown, but now there are also soybeans, malting barley, sugar beets, dry edible beans like pinto beans and navy beans, early-maturing corn, sunflowers, triticale, mustard seed, and canary seed.

North Dakota has 30 million acres of cropland and 12 million acres of rangeland. But it has a shorter season and more limited rainfall than richer Middle Western farm states like Iowa. Even so, in 1981 North Dakota had America's highest percentage of millionaires. On paper. Land in the Red River Valley was selling for $1,300 to $1,400 an acre. By the late 1980s it fell to a low of $650 to $700. It is now gradually climbing again.

Today North Dakota farming is a fairly high-tech industry with a high ratio of capital to labor, steady changes in methods, and a very high educational level; almost every farmer has spent time in college, and most have university degrees, a striking difference from some rural areas.

Prairie, down to about 700 people from about twice that in the 1930s (its official population of 761 is probably out of date), survives as a county seat, though its rival, Harvey, 18 miles northwest and better situated as to railroads, has done better as a marketing center. Even so, there is something exemplary about Prairie. It prides itself on being just 59 miles from Rugby, North Dakota, the geographical center of America. One is reminded of the address on the letter a girl in Thornton Wilder's *Our Town* gets from her minister when she is sick: "Jane Crofut; The Crofut Farm; Grover's Corners; Sutton County; New Hampshire; United States of America; Continent of North America; Western Hemisphere; the Earth; the Solar System; the Universe; the Mind of God."

Prairie itself has been the subject of several books. Margarethe Erdahl Shank called it Grand Prairie in her 1953 novel, *The Coffee Train*, about growing up in a Norwegian family in the 1920s. She described it:

> Grand Prairie was bounded on the north by wheat, bounded on the south by wheat; the sun rose on wheat, the sun set on wheat, and over

it all, by the Grace of God, blew the prevailing northwest wind. Almost anywhere you stood, in the little town of Grand Prairie, you could look up a street or down a street and see a field of grain in the distance, or in winter the rough-humped snow burying black loam waiting for the spring plow. . . .

Ranged about the little town stood houses whose every timber and all their furnishings had been wrested from a wheat field. Out of these houses, stooped old farmers, now retired, walked to the post office every day because there was really nowhere else to go. . . .

Larry Woiwode wrote a whole series of stories about Sykeston, his hometown, fictionalized as Hyatt, North Dakota, for *The New Yorker*. In his 1965 novel, *Beyond the Bedroom Wall*, and in his *The Neumiller Stories*, published in 1989, Woiwode imagines Sykeston, now a dying community, as it was in its heyday:

Every night when I'm not able to sleep . . . I try again to retrace the street. . . . It's the main street of Hyatt, North Dakota, and it's one block long. I lived in Hyatt from the time I was born until I was six and returned only once, at the age of eight, wearing a plaid jacket exactly like my brother's, too light for the weather, and ran up and down this street with changed friends, playing hide-and-seek between buildings that stand deserted, now that time has had its diminishing effects. . . .

The Town Hall is the heart of Hyatt—the sports arena, the theater, the polling place, the movie house, the dance hall, and the band building, the social retreat, the inoculation center, the courthouse, the village palace with its changing set of kings, and every Thursday evening during the winter, when a man from Fessenden wheels a wooden crate mounted on skate rollers into the building and seats himself on a stool like the stools used by shoe salesmen and straps on skates, the Town Hall is a noisy roller rink.

I come to the corner of the building and turn south on Main Street. At its far end I can see the raised roadbed and shining rails of the Soo Line, and beyond the rails, flanking and towering high above the Western store fronts of the street, four grain elevators. . . .

While the last of the pioneer settlers died in the 1950s and 1960s, many of their children survive and remember the old days well. Emma Bjornstad recalls how her father and an uncle brought their families out in a boxcar. Under the Homestead Act her father got 160 acres; when he quit farming, he had four times that. She goes on: "We were five sisters and one brother. We girls went barefoot all summer long. To save our shoes. When I got married, I moved 3 miles away and we bought 320 acres and we lived there for 25 years. When my father-in-law died, we bought his farm and then we bought another farm. I still have them. We farmed 47 years and Claus, my husband, dropped dead. We were on vacation. He was 65. He's been dead 15 years now.

"We had diversified farming. A little bit of everything. I worked outside with Claus just like a hired man. I'd worked with Papa. Harnessing horses, milking cows, cleaning the separator, that was nothing. I did the man's work. With my father, I plowed. Six horses on a gang plow. Papa had what he called a push binder. And Mama and I did the shocking. There was no money. It was during the Depression.

"When the threshing rigs came, I hauled bundles. But I didn't pitch. We'd drive the rack under a shock loader, and it would fill our rack. We were all girls, driving four racks. There'd be six men, three on each side of the separator, pitching the bundles in. That's the way we did it in 1920. The Norwegian and German-Russian women worked in the fields. Stacking hay. Mama and I cooked in a cook car for threshing. We had to carry water. We had to carry coal. We had to carry ashes. We had to bake bread, bake cake, bake pies. Working in the field was dirtier, but I got just as tired working in the cook car. I started that hard work when I was about 17."

Nowadays, Emma feels, when women drive tractors or combines, it's "just like sitting in the front seat of a car." She and Claus had just one child, a son, and he went into the navy and eventually became a captain. He has worked in Saudi Arabia since and in recent years takes African safaris. Life has changed beyond imagination.

Would she have changed things? "I tell you," Emma says, "you do work within reason and you're all right. If you're tired and played out, quit. I farmed my whole life. Sometimes I was so tired when I went to bed I couldn't sink deep enough to go to sleep. I was happiest when my husband was living and I had a lot of work to do."

Owen Rhys, whose father was among Prairie's few Welsh pioneer settlers, now a still hardy-looking man at 88, recalls how it was early this century. "When I was a kid," he says, "there was a farm over here and there was a farm down half a mile. And before that, there was a settler on pret'near every quarter section around here. When my dad started out in the 1890s, every quarter had a little log house or sod house on it. Claims, they called 'em. I can remember those boys from what became Hamberg hauling their grain into town. Before the Great Northern went in with their cutoff. They'd all go by here with their horses and wagons. I suppose it was in the early 1930s when the Model T trucks started coming in. Same with tractors. Those old International 15-30s. My dad bought his first one in twenty-six. I guess that's why my hearing is so bad. Because you run them and there was no muffler on 'em. The exhaust came out the side, you know, and just cracked out there."

I asked Gebhart Bauer, also in his 80s, what it was like to farm 70 years ago. "Well, Dad had two quarters and a 40," he began. "But 40 acres of it was lake, so you couldn't count that. It was roughly 320 acres, a good-sized farm in those days. We grew wheat, oats, barley, a little hay, alfalfa. We had cattle, about twenty-five hogs and poultry. Mother raised about a hundred for butchering and another hundred for laying. Sometimes she'd raise a few turkeys to market at Thanksgiving and have a few extra dollars.

"We were six boys and one girl, Maria. We had an eight-room white frame house. Four bedrooms. Big kitchen. We kept about twelve horses too. We always raised two or three colts. To replace the horses, you know.

"In the morning Father got up first. Never after five. Then he'd wake us up. To go to work. Mother got up and made breakfast. She had a wood-burning stove. Lit it with fresh wood and kindling. Dad never allowed her to soak a corncob with kerosene like some of them done. Us boys usually milked the cows. Mother milked too if we were all busy. The boys ran the separator. Yep, I've turned that thing many miles. You had to turn the crank 60 revolutions a minute. There was a little bell on that crank and when that bell quit ringing, you were going the right speed. If you got tired and slowed down a little bit, the bell would start ringing. We took the cream into town with horses and a buggy. Or by sled in the wintertime.

"We had a Model T. But you couldn't drive it on those dirt roads in bad weather. After twenty-eight Dad bought a Model A. That was a pretty nice car. Everybody could handle that. Even my mother drove that some.

"Once a week we had to churn butter. Usually every Saturday. We fed the skim milk to the calves and hogs. The hogs were also fed barley and oats. If we had a good corn crop, we'd feed them that too. We grew corn every year for cattle. Fed them the whole stalk, ear and all. Before breakfast we curried and harnessed the horses. We did a good hour of hard work before breakfast. Mother would water and feed the chickens. For breakfast she'd make eggs and ham or bacon that was homegrown, you know, and fried potatoes and toast, lots of bread. Mother baked eight or ten loaves twice a week. You burned up calories in those days.

"One plow was a three-horse outfit and the other was a five. We had a four-horse unit for cultivating and disking. You did a lot of disking in those days. Dragging was five horses on a 21-foot drag. Out in the fields, we'd probably rest the horses every other mile. If it was nice and cool the horses could go 3, 4 miles without a rest. If it was real hot, you'd maybe rest them every mile. 'Cause they really got to puffing. Twenty miles was a big day. In hot weather you'd probably do 18.

"You always went home for lunch to feed the horses. Water and feed them before you ate. You usually gave them a full hour to eat. So you took a rest. Not much because you had a lot of other things to do.

"My mother worked as hard. Oh, boy, she'd carry water from the well outside and put it in a boiler and put it on the stove. In 1925 we got a Maytag with a little engine on it that agitated the clothes. But before that she used a washboard. Rubbed out all the dirty clothing, you know, like our dirty overalls and dirty shirts; she rubbed them all by hand. Even after she had the machine, she didn't want to put them in dirty like that. She hung everything out on the clothesline back of the house.

"That was on Monday. Then a lot of times she'd bake something almost every day, like a cake or pie. She had a big garden. Good God, she used to can about 400 quarts of meat, chicken, pork, and beef, 200 2-quart jars. And probably that many quarts of vegetables. Beans, peas, corn, everything you can think of. Dad would never have been able to feed that big family without a big garden. We raised enough potatoes to eat and sell some too. We kept 'em in a root cellar below the house, just plain dirt. All that canned food was put down there on shelves. That cellar was always cool. Mother could take sweet cream and put it down there, and it'd still be sweet the next day.

"Us boys had to carry in water. A couple buckets three times a day. At night we did it for the next morning. Until we got electricity in 1924, we used kerosene lamps. Plumbing came much later. You'd just say, 'Well, I'm going out back,' or 'I gotta go.' Yeah, 'Clear the way 'cause I gotta go!'

"We went to bed as soon as the sun went down in the summertime. We were ready for sleep. Oh, in the wintertime when there wasn't that much work, then we'd have a little scrapping and raising hell, you know, and Dad had to come up and whip us a few times."

One of the Bauer brothers, Karl, stayed with us in Fargo when he went to college. Later he fought on Guadalcanal. After the war he went down to

Florida and got rich in home construction and real estate. On a visit to Russia in 1989, he got involved helping build six hundred houses for earthquake victims in Armenia. Old man Bauer, who sent Karl off to college with just $30 and a cardboard suitcase with a few clothes, sounds like a hard taskmaster; but like so many authoritarian farm family heads, he endowed his children with a strong work ethic.

During the Depression and Dust Bowl years, many of Prairie's farmers, no matter how hard they worked, couldn't pay their taxes and lost their farms. Janet Martin, 98, remembers it well. "It was terrible. The drought lasted 10 years. And grasshoppers. So many grasshoppers. The air was full of them. My Uncle John had 18 quarters of land. And you put all the seed out and paid the hired man and bought the gas for the tractor. You put in a lot of money. And got no return. Of 18 quarters, he saved just 7. One section he had to sell for taxes for $2,800. A whole section. I felt awful bad. We had no money. We couldn't save it. Yep, those were the years."

Gebhart Bauer feels 1934 was the worst year, though 1931 through 1936 were all bad really. "But 1934 was terrible. The Dust Bowl. I mean it just blew, and the dust got so bad you'd have to put lamps on to see inside the house. It got that bad. God, it was awful. We planted trees on the farm that year and carried water by hand. A 5-gallon bucket in each hole. Otherwise the tree woulda died right there. A lot of the windbreaks you see nowadays got planted then. And you can still see mounds in the fields where dust buried an old fence line."

I asked about grasshoppers. "A whole cloud of them would come down on a farm and eat everything in sight. You could walk on 'em three deep. Oh, sure, they'd jump on you. But they didn't bother you any though. They were big locusts, big fellows, with black in their wings. They flew almost like little sparrows. Alfalfa fields that had a little growth on them— maybe 6, 8 inches—when those grasshoppers finished, it was eaten right down to the ground. Nothing left."

Bauer was good at remembering changes in farm technology. He got his first tractor in 1936. "A John Deere. There is only two kinds. John Deere and all the others. I kept a team of horses for bad roads and winter. And feeding cattle. In the wintertime I hooked 'em up almost every day. We didn't have good roads yet, you know. I'd say it was 1955 before every-body had a good gravel road to his farm."

Rubber tires came in 1950. "The first tractors had steel wheels with big lugs on them. So after the war, rubber tires came in strong. And people cut the old rims off. We did. We took our tractor to the blacksmith and he cut the old wheel off and welded the rim on so you could put a tire on. Hundreds were done here in Hawk County. And people tore out some of their fences to make bigger fields."

He first took money from the government in 1946, though farm programs started back in the 1930s under Roosevelt. "We got some money for planting sweet clover to enrich the soil and tie it down so the wind didn't blow it away. They had programs for topsoil working, trashy fallow, you know, when you left the stubble on top. The checks weren't so very large, $400, $500, but it was a nice piece of change."

In the 1920s 15 bushels per acre was an average wheat yield; 25 bushels, a bumper crop. In the 1940s, some crops went over 30 and prices held up in relation to inputs. Today wheat yields around Prairie run 30 to 45 bushels an acre; a few crops have gone over 50 in the past 10 years. Average yields are about double what they were 60 years ago. New rust- and disease-resistant wheat varieties with shorter growing seasons make the difference, as does chemical fertilizer, first introduced locally in the 1940s but not used universally until the 1960s. Anhydrous ammonia is the main fertilizer applied.

Prairie itself reflects every change in the farming around it. Indeed, the town is so much a part of the surrounding countryside, I often found it startling, going back in recent years, to open the door in the morning to go outside and once again be overwhelmed by the flat brightness. The fierceness of the light, the enormity of the blue prairie sky, the perfectly flat, unbroken plain visible between the houses and down every street; I've never gotten used to it. Remember Carol Kennicott's first impression, from the train, of Gopher Prairie in *Main Street*? Sinclair Lewis catches the sense of it very well:

> And she saw that Gopher Prairie was merely an enlargement of all the hamlets which they had been passing. . . . The huddled low houses broke the plains scarcely more than would a hazel thicket. The fields swept up to it, past it. It was unprotected and unprotecting; there was no dignity in it nor any hope of greatness. Only the tall red grain-elevator and a few tinny church-steeples rose from the mass. It was a frontier camp. It was not a place to live in, not possibly, not conceivably.

Like Mrs. Shank in *The Coffee Train*, Lewis emphasized how the town is almost overwhelmed by the prairie:

Main Street with its two-story brick shops, its story-and-a-half wooden residences, its muddy expanse from concrete walk to walk, its huddle of Fords and lumber-wagons, was too small to absorb her. The broad, straight, unenticing gashes of the streets let in the grasping prairie on every side. She realized the vastness and emptiness of the land.

Sauk Center, Minnesota, on which Lewis based Gopher Prairie, is only about 40 miles from the North Dakota border. As Carol tours Main Street, he writes, it could be any of "ten thousand towns from Albany to San Diego." Lewis is gleefully disdainful:

A small wooden motion-picture theater called "The Rosebud Movie Palace." Lithographs announcing a film called "Fatty in Love." . . . Dahl & Oleson's Meat Market—a reek of blood. In front of the saloons, farmwives sitting on the seats of wagons, waiting for their husbands to become drunk and ready to start home.

The Ford Garage and the Buick Garage. . . . The most energetic and vital places in town. . . .

Ed James, 76, recalls how Prairie's Main Street looked in its heyday: "In the old days, there was Bergstrom's Department Store. That was like a city emporium with those whirring wire baskets that swing from moving cables over the counters. They carried money and receipts back to old Mr. Bergstrom, who sat on a dais in the middle of the store. Bergstrom's had everything, groceries, hardware, clothes. A sales clerk would yank on a wooden-handled rope to pull a basket down, and another to get it back up. It was a big two-story red brick building, like the hotel and the auditorium, and with its first-story front of clear glass, Bergstrom's was really something way out on the prairie. After it you had Anderson's Jewelry, Gamble's hardware store, the funeral parlor upstairs, and the auditorium where they had the movies and minstrel shows.

"The circus used to set up two blocks west of Main Street. They came by train. They'd have a parade, march into town, with an eight- or ten-piece band, steam calliope, elephants, cages, tents. It was just a fly-by-night deal. They'd come in for a day or two and then they'd go. We used to have to go around and peddle handbills house to house to advertise the movie. Us kids. And then on Saturday night we had a cowbell, and we'd walk around the block there, ringing the bell and hollering, 'Show starts in 15 minutes!

Show's about to start!' It was my dad that ran it, so I didn't get paid or nothing.

"Talkies came in 1928 and they had a turntable underneath the projection booth. The show would start and you'd set the needle on the record disc while you were watching the screen. I had to do it for my dad lots of times. If you didn't match the sound just right, the crowd started whistling and hollering. The worst was the chariot race in *Ben Hur*.

"There was a barber shop in the auditorium too. Then came a meat market—earlier it was a confectionary—and Swan's Store. On the other side there was Sam Levy's men's store, Ed Peterson's grocery, the Prairie Cafe that caught fire in '29, and a pool hall. They made that into a post office later, and there was a little cream station next to that. And then there was a vacant lot where the bandstand was. We had concerts Saturday nights, right downtown there in Prairie. I played the clarinet. Then came Fisher's Drug Store—the only drugstore in town—and a shoe repair shop, the First National Bank building, and another bank on the corner that went bust and it became a pool hall. Across the street there were lumber yards on both sides, right up to the railroad tracks, and at the other end of Main Street, Riley's Hotel.

"I worked in Swan's Store there for quite a few years. Used to deliver groceries with an old Chevy. We put a box in the back end of it. Most families had ice boxes. When it was time to take groceries, about five o'clock when I got out of high school, you'd just drive around and you'd take this one and run in with the groceries and put 'em on the kitchen table. It was a service all the stores in town provided. Another job I had was breaking up the wooden cartons that the groceries came in. They had a bin in the back and they threw all the boxes in, and once a month or so I'd have to go in there and chop 'em all up. It was mostly dirt roads in town then. Tire tracks and ruts made by the wheels. Oh, there'd be terrible ruts in those roads when it got wet. I think the ruts just stayed. They didn't grade 'em too much."

Karl Fischer, a retired banker, remembers the Prairie that James describes but thinks the old ethics are gone for good. "I remember one time when I was a kid and they were giving out oranges downtown. I ran home and told my folks that we could get oranges. And my dad says, 'I don't want you near that truck. We're not going to take a handout from anybody.' The philosophy of people at that time was so different. If you had all you wanted to eat, if you could pay your bills, if you kept your house warm in the wintertime and enjoyed your neighbors, why, what else was there?

Then all of a sudden there was this trace of greed, you know. People started to make money and then they wanted more money and, finally, money kinda took over."

Prairie's people, trying to come to grips with the present crisis, often compare it to the Depression. Many say that in the 1930s if a farmer didn't make his payment, he lost his land. Now he can probably go to somebody else to bail him out. Another change is that farmers now need more land. Some say they need at least three quarters—640 acres—to make expenses. Anything over that, they say, they are "working for themselves." If you need 640 acres to break even in North Dakota, it is easy to see why so many farmers farm 1,000, 2,000, 3,000 acres or more. The more you farm, the more the profit. It is like putting up an apartment house. If you need ten units to pay expenses, you'll try to have twenty. Farming is the same.

The 2 years of drought knocked such calculations sideways for many. Rain is everything on the Great Plains. There is never enough. But while total precipitation in Hawk County was nearly 19 inches in 1986 and over 15 inches in 1987, it fell to 6.8 inches for the whole of 1988. It was not until the fall of 1989 that Prairie, in 6 weeks, got more rain than in the previous 2 years.

A normal wheat crop around Prairie is 30 bushels an acre. Local farmers say they can live with any price over $4 a bushel. In fact, the way it used to be set up, government support prices would have brought about $5.75 a bushel in the fall of 1989. Instead, with the government trying to reduce farm subsidies, it dropped to $3.35.

Most of the farmers I met around Prairie felt they had lost half to two-thirds of their wheat crops when the rainfall failed in 1988 and 1989. Yields fell to 10 to 15 bushels an acre both years. What saved everybody then were federal crop insurance, which covered 3 or 4 bushels an acre, and disaster payments, covering another 4 or 5 bushels. Wheat prices temporarily went up, so durum was selling for $6.80 a bushel. Many sold whatever they had stored in their bins. It was commonly felt that when farmers did go broke—and several in Prairie did—they were already in the "hurt bag" and the two successive droughts were just the last straw. Still, almost everybody suffered. Whereas in a normal year a farmer could figure on getting about $120 an acre, in both 1988 and 1989, even with government payments, he got just $75 or so. The farmers I know best generally felt they had lost about a third of their incomes both years. All said the pinch came when they had to keep up payments on their land and debts owed for equipment. Some feared that without several good years in a row, their real troubles lay ahead.

A few farmers, by emptying their bins in 1988 and selling at peak prices, like the $6.80 a bushel for durum wheat I mentioned, actually did better than in a normal year, though they couldn't repeat it in 1989. Jeff Muhlenthaler, a schoolteacher who also farms west of Prairie, told me his yield in 1988 was just 10 bushels an acre, down from a normal 35 bushels. Even so, he was not hurting. "In fact," Muhlenthaler explained, "for the last 5 years I had to borrow money to put the crop in. After the drought I didn't." Farmers usually keep some wheat—or carryover, as they call it—in their storage bins, partly as insurance for a rainy day but mainly in hopes the price will come up. What Muhlenthaler did was sell all the grain in his bins. So with crop insurance and federal disaster relief payments, along with the high price he got for the grain he sold, he actually came out better than he did in an average year. But it was a one-time shot. As he said, "Looking 2 or 3 years down the road, I don't have any carryover grain, and then what?"

There were enough farmers like Jeff Muhlenthaler across the Middle West that some outside observers talked about recovery. Paradoxically, in this view, the finances of farmers seemed to have been restored by the worst drought in more than half a century. And you could say that farmers benefited in two ways. They got disaster relief from the federal government on crops that failed in 1988 and 1989, and higher prices on the Chicago markets on the crops that grew or they had stored in grain bins. Average real net incomes reached a new high in 1989. At $40 billion, farm exports in fiscal 1989 were 50 percent above fiscal 1986's $26 billion.

But as farmers like Jeff Muhlenthaler were the first to say, reserve stocks of corn and wheat, after most of them were sold in 1988–89, were unusually low. A short-term advantage was that farmers did not need to fear a price collapse if they produced a bumper harvest in 1990 or 1991. And after a long decline in the price of farmland, after 1988 it seemed to be going up again. So the recovery could prove very illusory. What was needed, all Prairie's farmers agreed, were 2 or 3 good years and a chance to build up their reserve grain stocks.

It is said Hawk County has more large farm equipment than any other county in North Dakota. If so, after the drought years, its debts are probably bigger too. A large tractor now costs $60,000; years ago you could buy two or three farms for that. To pay for something so expensive on time means a farmer has a lot of interest to pay and he can't take many bad years in a row. Once it wouldn't have mattered so much. If hard times came 50 years ago, a farmer still had a place to live. Today if he knows he can't pay his debts on land, equipment, even at the grocery store, he'll probably have an auction sale and maybe sell off his land and equipment and pay off his creditors and still come out with a little money. Then he will either get a job and move to town or help somebody else to farm.

As mentioned, you don't see many people in the fields anymore in the Middle West. This is one of the most striking differences from the Third World, where every rural landscape is dotted with people plowing, hoeing, harvesting, or going along the paths and roads. Fischer, the banker, told me how, before he retired, he tried to get one of the Johnson chain of stores to open up an outlet in Prairie. "Three men came down one February," he said. "We'd had about 3 inches of snow on the ground. They came down in the morning, and we agreed to have lunch together. At noon we walked over to the hotel and sat down. And one of them said, 'You know, I'm really puzzled. We drove out into the country, and we couldn't see any tracks in the snow into those farmyards.' 'Well,' I said, 'some of them are in Arizona, some in California, some probably aren't even up yet.' He asked, 'Don't they have any livestock?' I said, 'No, they don't. We've got a lot of grain farmers here.' Well, before they left they told me, 'You know our stores cater to farm families. That need things for their cattle or farm-type clothes, things like that. We just couldn't find anybody. All these people, they aren't around. I don't know how many miles we put on, but we drove all over and we didn't see any livestock.' And that just about tells the story about Prairie."

It is the same story all over the Middle West for so many grain farmers. Winter is a time for holidays.

PRAIRIE'S MAIN STREET in 1990, with its two-story brick shops, half of them empty, and its huddle of cars and pickup trucks, looks like so much of small-town America today, just holding its own. The residential streets, with their trim houses, elms, maples, and oaks and carefully tended lawns and flower gardens, look pleasant and lasting. It is Main Street that provides the real barometer of decline.

At the south end, opposite the Standard station and Prairie's only grocery store—quite a large Jack and Jill supermarket—is the old Riley

Hotel, three storied and red brick. When it was built in 1905, it was Prairie's most imposing building. It stopped taking guests years ago, but Mabel Rowlands, the owner, and her two grown sons still live there, and Mabel keeps the downstairs restaurant going. With its linoleum-covered floor, a long counter on one side and a row of booths on the other and tables covered with yellow oilcloth in between, the restaurant hasn't changed much in the lifetimes of its customers. Mabel cooks a homemade dinner every noon—meat loaf, mashed potatoes and gravy, maybe string beans or corn on the side, some days baked ham or spaghetti or fried chicken, and always a roast on Sundays. Every Saturday Mable bakes cinnamon rolls, and Sunday morning the Baptists, Catholics, and Lutherans come in for coffee after church. But the hotel lobby is bare now, its proud old mirrors and plate glass windows dusty and flyspecked. A handmade sign, "J & L Sports—Hunting, Fishing, Archery," is a reminder of the time the two Rowlands boys tried to get a sports store going in the basement. Outside the hotel a sign, "Bus Stop," is a reminder a minibus will stop at 11:00 A.M. each day to take a few passengers into Jamestown on the big east-west highway, where they can transfer to a Greyhound into Fargo. In the old days several passenger trains stopped in Prairie every day; it was a comfortable overnight trip in a Pullman to Minneapolis; now the little bus is Prairie's only public transport to the outer world.

Going north down Main Street on the east side, back on the west, there is the usual row of stores one finds in small town America:

Peterson's Men's and Boys' Clothing, now mainly work clothes. In 1989 it merged with a store next door run by a farmer's wife who sells carpets, gifts, and real and artificial house plants.

The local office of the Herald-Press, which used to be two weekly papers, now combined. A faded banner in the window hails the high school team, "Welcome Back Orioles." Partitioned off to one side is the office of the Commodity Credit Union.

The post office, a partition of glass and brass shutting off its back room. A writing stand is against a wall hung with official notices and "Most Wanted" posters.

The First National Bank. A handsome stone-and-brick facade adorned with boxes of red geraniums, a patch of grass, and a rotating clock and thermometer, much consulted and commented upon by passersby.

Fisher's Drug Store, surprisingly well stocked, where you can buy *The Wall Street Journal* and daily papers from Fargo and Minot. The sense of a going thing dims when the druggist tells you he plans to retire next year and will close the store if he can't find a buyer. Prairie's only doctor says this would leave his elderly patients without any way to fill their prescriptions.

After a law office and an insurance agency, what was Larue's Jewelry and Sports Shop is now empty. Across the road, extending to the railroad tracks, is a lumberyard. The owner says he needs to supply materials for the construction of at least three new houses a year to stay in business. Last year he had none.

Across the railroad tracks the Thrift Shop displays piles of cast-off clothing and scratched and broken toys. The taxidermist's shop beside it is closed and empty inside, as is a blue-painted frame house with a "For Sale" sign on the porch. By the Soo tracks beneath a clump of cottonwood trees on Main Street's western side is an old Rock Island tractor with steel wheels, parts of two other cannibalized tractors, a trailer full of car parts, and a red Chevy pickup with the front end smashed in. These go with Steve's Body Shop, one of Prairie's few new enterprises. Beyond it is the medical center, where Prairie's last doctor holds office hours.

South on Main Street's west side is an empty building that long ago was Swan's Grocery Store. A sign in the window, "Stensrud Electric—Zenith Color Television," identifies its last former occupant. Next door is the "Big A" restaurant, going strong with its bar and bowling alley. Outside are Pepsi and Coke machines, both in operation.

Then comes the old auditorium, brick like the hotel, nowadays closed except for special performances. It used to be Prairie's movie theater, going back to the 1920s, and if you peer into its locked double doors, in the gloom you can just make out the faces of the film stars that adorned its walls. There are Robert Taylor and Vivien Leigh, W. C. Fields, Boris Karloff as Frankenstein's monster, Bob Hope and Bing Crosby in Arab outfits on the road to somewhere, Rita Hayworth in slinky black velvet, John Wayne in a cowboy hat.

Next door are a barber shop, an insurance agency, the Gamble's Hardware Store with an undertaker upstairs, the Checkerbox Boutique with women's clothes, and Prairie's third big brick building, the old Bergstrom's store. Its upstairs, once divided into offices and apartments, is now empty, as is the cavernous store itself.

Away from Main Street, but part of the business of the town, are the Prairie Motel and, a short distance down the highway, the Cabaret, a large makeshift building which is noisy and crowded from late afternoon on, with its bar, restaurant, and game room with video games and a pool table.

There are dances and bingo Saturday nights, and you can get a 12-ounce choice top round sirloin, cole slaw, a baked potato, and wine, even if the house red turns out to be pink and sweet. (Wine tasting like cherry pop seems to be popular throughout the rural Middle West.) Also on the highway there is a parts shop, an implement repair shop with a lot of farm equipment lying about, and a garage with its "Wash 'n Wax 25 Cents Self Service." In summer the Filling Station opens, offering burgers and shakes and a killer called a Boston shake, sort of a combined hot fudge sundae and malted milkshake.

The largest congregation in Prairie's four churches is Lutheran, followed by the Catholics, Baptists, and Congregationalists. Scattered around town are also a legion hall, a Masonic lodge, and a town park with a swimming pool. Contradictory signs at the park's entrance say "Camping" and "Park Closed 12 Midnite Thru 6 a.m." As almost everywhere in Prairie, its grass is neatly mowed and there are lots of flowers in summer— marigolds, nasturtiums, bachelor's buttons, gladioli, mums, zinnias, and hollyhocks.

A greenish blue water tower rises over the park, but even higher are the gray roofs of Prairie's enormous grain elevator. Its structure dominates the skyline, and up close the elevator hums and sisses and makes faint plunkety-plunk sounds, like it really was the town's heart. Since it consolidated, the elevator handles grain for all Hawk County. But Prairie's gain in jobs is offset by a loss of them in all the small communities around; they now need only one or two people to run their satellite elevators, as most grain gets trucked in to Prairie.

What really keeps Prairie going is its standing as county seat. North Dakota is hard-pressed to keep its fifty-three county governments going. If it ever loses its Hawk County Courthouse, with its visiting judges and law officers and clerks, Prairie would just about fold up. The courthouse is an imposing red-brick Victorian building, with turrets and gingerbread cornices. Built in 1895, it was sandblasted a few years ago and looks practically new. The sheriff's office and county jail are in the basement.

Across from the courthouse are the wooden, barrackslike, one-story government offices of the Farmers Home Administration (FmHA), which makes loans; the Agricultural Stabilization and Conservation Service (ASCS), to administer crop supports and other farm programs; the state university's extension service; and the county's social services office.

East of the courthouse is Second Street N.E., known to everybody as Quality Hill. It is Prairie's one grand street, about three blocks long and almost overhung by the branches of stately old trees. Here a few old Victorian mansions—one of them even has a ballroom—sit solid and comfortable behind hedges and fences; all are lived in and kept up.

Another thing that sets Prairie apart from most little towns is its fairgrounds. Just south of town, with an entrance gate on the highway, it is like another little town: freshly painted white wooden buildings, barns up on a slope at the southern end, big bleachers and grandstand facing the race course, and behind it a midway partly shaded by clumps of cottonwood trees. The fairgrounds is my favorite place in Prairie; to the east, west, and south it looks out on open countryside, and you can see for miles; when in town, I usually jog there before breakfast.

A sign on the big stone entrance gate reads, "Hawk Co. Fair—Capital of Thoroughbred Horse Racing in No. Dak.—Prairie, N.D.—June 24–28." Most of the jockeys are Indians—Sioux from Devil's Lake or Fort Totten, or Chippewa from White Earth in Minnesota. The 4-H pavilion and home economics hall have been there at least since the 1920s. The big stone dance hall and roller rink was built during the Depression by the WPA. The Hawk County Museum is housed in what used to be a two-room schoolhouse. But the two big sheds full of old farm machinery are new. The Catholics rebuilt their food stand in 1987, and the betting area was set up when it was legalized in 1988. But the beer garden under the grandstand and the three big cattle and horse barns and the long shed for poultry go back years and years. In summer the county fair is still the biggest event for miles and miles, even if it lasts just 4 days.

WHENEVER I VISIT Prairie, I always figure the place to start is with Charles Pritchett, the editor and owner of the *Herald-Press*. Charlie, as everybody calls him, came out to Prairie to farm in 1975 when his wife, Mary, inherited 800 acres from her father; it had been in the family as an original homestead on the Soo Line. As a city boy who spent 16 years after college working as a chemist for the 3M Company in Minneapolis, Charlie sees Prairie with a mixture of hometown pride and an outsider's detachment.

Charlie stuck to farming 8 years, "endured" he now says because he wasn't making any money. When the chance came to buy the weekly papers of Prairie and Harvey and combine them into the single *Herald-Press,* he jumped at it. "I got a thrill out of seeing stubble turning into beautiful black dirt behind the digger," he told *The Minneapolis Star and Tribune* in a 1987 interview. "But when I got to thinking about costs and prices, I managed to isolate my emotions."

Charlie was still farming part-time when I went back to North Dakota in 1982 for the first time in 34 years to do research on my family's life there for a book. With great good luck, I found he had the bound copies of the old paper for 1925 to 1932, the years my family lived in Prairie, just sitting on a shelf in his office. Without further ado, Charlie let me and my research assistant set up shop at a big table in the back room of his newspaper office; a funeral parlor across the street loaned us black and maroon folding chairs. Since it took us all summer to go through the papers, I am really indebted to Charlie.

Tall, black-haired, always on the move, and with a sudden, wide grin which spreads over his whole face, Charlie looks much younger than his 54. So does Mary Jane, his pretty, fair-haired wife; it is hard to believe this handsome pair has six grown children. Just a married son, Jack, is left in Prairie.

"People are hanging on," was Charlie's general verdict on my visit. Many, if not most, of Prairie's farmers are in permanent financial straits for the foreseeable future, Charlie feels. Even with the 1988 and 1989 droughts, as I've noted, not too many farmers went broke and out of farming entirely, as they would have during the Depression.

"It's hard times," says Charlie. "There have been distress sales. Mainly people work out a deal with the Bank or FmHA or the Federal Land Bank or whatever. They just work out a deal. Whoever's holding their mortgages gets what they can get. And they write off a loan. The bank in Prairie was very solid. Whatever loan problem it got into, it got out early enough; it didn't let people get hit. Then 1988 was an election year and, after the drought, the government pumped millions and millions into the state. Row crops like sunflowers, beans, and corn have been good this year. Small grains have been erratic. Some farmers lost half their wheat crop or more in 1989. At the same time, land values are up 5 to 10 percent, which is a welcome turn."

Charlie has a theory that timing—when you happen to get into farming—is everything. "It's unsupported by any research at all," he says. "I think, by an accident of demographics and history and pattern of settlement, that a large number of landowners around here, and maybe in North Dakota, got into farming right after World War II. The middle to late 1940s. For the next 30 years anybody who could keep equipment running could make money in farming. And it was capped off in the 1970s by a boom. Everything was being farmed, so there wasn't a whole lot of room for new people to get in and get into trouble. And land values went up fourfold in those 30 years, from $150 to around $600 an acre by 1980. It fell in the 1980s to under $350 an acre and is about $400 to $450 now."

I told Charlie it was like the words in *Ecclesiastes,* "Time and change happeneth to all."

"Yeah, so by and large farmers were resistant to crisis when it came. I mean they had money in the bank. They could take some of their CDs and spend it for seed and fuel and chemicals and a new tractor and remodel their house and take a trip to Florida and still be operating out of their own war chest. Whereas people who were just getting into farming had to go to the bank and borrow money at 18 percent or 20½ percent or whatever it peaked out at. The guys who were using their own funds, they just never really got in a hole. I really think it was the over-20-percent interest rates that were killing. I'm still paying those loans off. I'll never live long enough to pay them. It'll go about 20 years after I'm dead. Now the interest rates are down to 11 to 12 percent. But the loans don't get paid off. The principal just gets refinanced."

We were sitting in the old *Herald-Press* office on Main Street, with its shelves of bound yellowing newspapers going back to the nineteenth century. A stuffed fox, left for a customer by Charlie's son Jack, who did taxidermy on the side, stood poised and lifelike on a counter. The office was a comfortable sort of place, where people came in to pay a bill or place an ad; the editorial office and printing press were in Harvey. Charlie said he was going to let the office go; Mary would handle everything from a desk and counter in a corner of the store next door. I was sorry to hear it. It was one more thing to lose.

I asked Charlie how a farm in Prairie went from one generation to the next. He said a lot of farmers weren't willing to let go, especially if they went through the Depression. "Instead of making an orderly transition," he said, "which means letting go of control, they'll hang on and hang on until they're dead. And then they lose it all in estate tax. If it's not a husband or wife, you lose a heck of a lot. It just depends on how big an estate is. You almost have to sell a third of a farm to keep the other two-thirds, which isn't at all what the dead person had in mind. They'd kept saying, 'Some day this will all be yours.' "

A lot of those who held on to the bitter end did seem to retire and rent their land. Somebody had told me two-thirds of the land around Prairie was owned by absentee landlords. I asked Charlie about it.

"I'd be surprised if it's that high," he said. "But I'd believe it. They're not necessarily absentees in the sense of being in Arizona. A lot of them live right here. It's so little talked about because generally the people are fairly powerful or influential people in the community. So one-third of Hawk County's wealth is going to Arizona or somewhere. That's a reality. You don't hear much about it because it's so common.

"All I know is, it really hurts when you see a guy coming in driving his Buick Town Car with a sticker on it from Ajax Auto Sales in Scottsdale, Arizona. The clubs he uses out at the golf course, he got at the pro shop in Scottsdale. You know, they come here in the summer and they live in their house and go back in the fall to Arizona or Texas or Florida and spend their rent money from the farm. They're very loyal to Prairie. So many people who are loyal to little towns haven't worked out the fact they're the ones helping to do them in. And are very resentful at even the suggestion. They say, 'I've made that money honestly. It's my money. It's nobody's damn business how I spend it.' They're happy and contented and having a good time. Headed for the golf tournament. And they're by and large insensitive to the plight of the young farmers."

It is true. You don't have to be in Prairie very long to see the gap between the richer older generation and the poorer younger one. The old are either retired farmers who rent their land or grain farmers who do what people joke about as a wheat-Arizona-beans-Hawaii rotation. The younger farmers almost invariably seem to live in trailers, renting what land is available, their wives working hard.

Charlie agrees. "Here again it's judgmental. But I just see it so often, and it just breaks my heart. These older farmers who made it in that little window of history from 1945 to 1975 when anybody could make it. They really think it was because of their management skills and their brilliance and their courage and their all-of-that. There's just so little sympathy for people who made the same decisions on the same basis in 1980 as these guys made in 1950, but it went sour. It went out of control and turned against them. I can't imagine you'd find any young farmers here who aren't under stress."

As I went around Prairie, I was to find a good many farm families are anxious to transmit the land from one generation to another, to see their children go into farming. A lot depends on the parents' generosity. No other occupation in America has such father-to-son continuity. Not all fathers, however generous, have big enough farms to support two families. Charlie told me, "I've heard about kids who really like farming but their fathers tell them, 'You can't. I won't let you do it to yourself. Get an education. Find a job somewhere else. If in 10 years things look better and I can see my way to retire, then you can come in and I'll be glad and we'll try to make room for you. But for now, get lost.' " Others, sometimes rich, with or without sons, retire but hold on to their land, renting it, as Charlie said, as long as they live.

Most of the younger farmers I was to meet rented their land, borrowed, and were heavily in debt; their wives either had full-time off-farm jobs or worked in the fields driving machinery, virtually extra hands. Cash rent in Prairie in 1990 was $30 to $35 an acre, half paid in advance, half at harvest. During the drought, a few Prairie landowners demanded the whole year's rent to be paid up front; in one case the renter was a son-in-law, and in another a nephew; both owners involved were very much a part of Prairie's Arizona-wintering richer older generation. The drought also made sharecropping less popular with landlords. Where it is still done, a two-to-one split is most common, with the renter paying for all the inputs such as fertilizer, pesticides, fuel, and machinery, and the owner just paying taxes. A few have a fifty-fifty split, with tenant and owner equally sharing a farm's expenses and income.

Far more young people in Prairie want to stay home and farm or make their livelihood some way in town than end up being able to afford to do it. This is a crucial point about American rural culture I want to make. City people the world over often do not see why anyone would prefer to live in a rural setting (excepting the ideal of the country house within reasonable driving distance of a city center). In Latin Catholic culture, the bright lights are always in Manila or Rio or Mexico City. This may be related to Latin culture's Spanish and Portuguese feudal origins. But in societies like Egypt or Java or India, village culture is often on a higher plane than terrible urban slums; villagers forced to work in cities often feel their "real life" begins when they get home.

Generally speaking, I found this sense of attachment to one's native place very strong in Prairie, and also among Crow Creek's farming families. Further evidence this is true of most American rural people came in the agriculturally prosperous 1970s, when we experienced the start of what proved to be a short-lived reruralization.

So it was poignant to hear Charlie say that about 95 percent of Prairie's high school graduates go away to live, as many as 90 percent of them first attending college. "I was chaperoning a dance at the school a couple of months ago," Charlie told me, "and had time to kill and was just browsing through the halls looking at pictures of the graduating classes from the last

10 years. And I've lived here 15 years now and known most of these kids. And few of them have stuck around."

"What is being lost?" I asked Charlie. "When we're losing so many farms and small towns, and all the schools and churches and young people with them, isn't something culturally important to America being lost?"

Charlie replied, "Maybe people are going to say: I don't see anything being lost so far. Well, something is, though. . . ." He thought for a minute and then went on. "*People* are being lost. The culture is disappearing. The out-migration. The number of us here diminishes, diminishes, diminishes. Kids that were raised here have gone to Bismarck or Chicago or Los Angeles. Now, I got six kids. Jack will be here. But my other children's children will be born in a big town somewhere."

He recalled, when he met Mary in the early 1960s, Prairie had close to 1,000 people. "Now you gotta stretch to find 750. It may be down to 700. You go around the county and there's no remaining trace of farms that were there 20 years ago. Just trees, waving their branches, where houses used to be."

Harvey's population went up several hundred in the 1970s and 1980s as more farmers retired and moved to town. But its school enrollment has dropped. With ten fewer children a year at, say, $1,500 per pupil in state aid, that is $15,000, the pay of a teacher.

"Oh, the Twin Cities are booming," Charlie said, explaining he had worked in Minneapolis and was born in St. Paul, the nearest really big cities to Prairie. "I get there many times a year. I have two daughters living there. Fargo's doing well too. All of North Dakota's urban counties are gaining people and getting richer. All its farming counties are losing people and getting poorer. And just because I live here doesn't mean I think this is necessarily a better place to live. Or that somehow culturally, spiritually, it's superior. I don't hold to that. I don't take the dim view of metropolitan living that most of the people here who have never lived there do. St. Paul, where I grew up, is really a series of little self-sufficient communities where families can live and grow and expand. And you've got access to the arts and to sports events, so many things. We still go down there a lot. We've got a lake cabin down there in Wisconsin where we'll retire some day.

"Oh, if this whole thing went, I would care. Very much. Rural North Dakota offers a lifestyle you can't find in the Twin Cities. Whether it's a better lifestyle. . . . You give up something for everything you get. I just think the country would be definitely poorer for not having rural life as an option. I could go right back and live right where I lived in the heart of St. Paul. I'd miss a lot of this. But I'd enjoy a lot of that too."

Charlie was voicing my own sentiments. My point is not that urban or rural life is better or worse—both have advantages. My point is that urban culture *requires* a rural base, the foundation of its basic institutions like the family. The idea is not to make value judgments about what is good or bad, but rather to define what works and what doesn't work. And unless there are enough Prairies, American urban society is not going to work.

Charlie agrees that after years and years of steady decline, little towns like Prairie are now faced with a severe crisis. "A lot of these little towns are approaching it. It may take 10 years or 20 years to go over the brink, the brink where it can no longer sustain Main Street businesses. Look at the history of Hamberg. Or the history of Manfred, right up the road here. Or the history of Bowdon and Chaseley and Sykeston and Hurdsfield." Charlie named even-smaller communities around Prairie which were becoming ghost towns. "These little towns were thriving 30, 40 years ago. They had two or three banks, a couple of hardware stores, a couple grain elevators, clothing stores, blacksmith shops. It was because there were a couple of farms on every section. And every farm had four or five kids.

"Okay, it was a short-lived era for Prairie, 1880s to 1980s. I agree, 50 years to grow, 50 years to decline. The pattern is that, one by one, businesses start to drop off. And pretty soon there isn't enough to bring people in town for *anything*. So when they go to the bigger towns like Harvey or Minot, they get everything cheaper. And my fear is that Prairie is a town that's getting close to that. John Deere is out, gone. We've only got one implement dealer left."

To arrest Prairie's decline, Charlie said, its people needed to shop in town. "If this quality of life is something they treasure so much, it's worth paying $22 to buy a pair of slacks here that you might get in Harvey for $17. It's worth paying a little more. As long as people keep going to the bigger towns and say, 'Well, hell, I'm not going to pay that extra 5 bucks,' fewer and fewer small towns are going to survive."

Charlie said a contrary and hopeful, if hard-to-explain, phenomenon in Prairie is the way most of its remaining stores up and down Main Street have been taken over by fairly young people, mainly college graduates, with small children. Where are they coming from? Charlie says mainly from other rural areas in North Dakota. He thinks many are children of farmers who didn't have a way to go into farming at home but didn't want to go too far away.

Prairie's 4-year high school, which began the 1980s with 110 pupils, ended the decade with only 58. But its primary school, with 89 in 1980, had 142 in 1990. One sees the same thing at church. A few years ago the Sunday congregations were conspicuously elderly. Now they are overrun with young parents and small children. What explains it? Charlie admits to being baffled. My guess is they want to continue living in small-town North Dakota and they feel Prairie has a chance of making it.

IN ALL RAPIDLY changing agricultural communities—one sees it in Third World villages too—there are a few stubborn individuals who try to stop the clock. It never works. New technology remorselessly increases production, and the accumulation of more wheat or corn or whatever it is lowers prices for everybody. Yet one still finds the farmer—there will be an example in Crow Creek too—who tries to isolate his farm enough from the local economy so that he and his family can keep right on going the old-fashioned way with cows, pigs, chickens, a big garden, lots of canning, and very old, much-repaired machinery. I suspect one can find a few in every farm community.

Culturally they matter because they are so traditional; they show us how an institution like the family used to be. To drive into the Alfred Zuckerman place south of Prairie is like going back 40 years in time. There is a '49 Ford pickup in the shed and an International 1066 tractor just about as old. There are three milk cows in the pasture. Chickens and hens strut about.

"How can a farm like yours survive?" I asked Alfred Zuckerman, a red-faced, cheerful-looking man of 73. "We have so far," he replied. "I'm sure it can be done." Zuckerman said he farms 480 acres, bought from his father. "We own what we got and we don't rent from anyone. You go renting land, you gotta borrow money to pay for it. If you don't get it back, you're in debt."

He has never used chemical fertilizer. "Nope, long as I'm farming. Yield is good enough. 'Cause we've got shallow wells, yup. They're a little less than 30 feet. The neighbors got over 100 feet. You've got a bad pocket here. Never used fertilizer. Nope. Never have. Nothing! Just good summer fallow. Keep it black for 1 year. I'd match my crops, dollar for dollar, with any of my neighbors. I think I would. I don't get as many bushels as the ones that use fertilizer, but I'll end up with as many dollars in my pocket. I have no expenses. And I don't take credit either. Don't borrow on my crops."

I was a little surprised to find Zuckerman is in the government program for crop supports on his wheat and oats. He is pleased with his 160-acre wheat base for his three quarters. Like any old-fashioned farm that grows its own oats and feed, he does not plant that much wheat. If he doesn't completely escape all the paperwork that comes with the many federal regulations on crop rotations and percentages of land set aside to lie fallow and fluctuating loan rates, he does avoid the bureaucratic procedures more than most.

A day on Zuckerman's farm starts early as he and his wife milk the cows before breakfast. "We usually milk again at five o'clock in the afternoon. We have our supper and then we give the calves hay. Bed 'em down. Around seven o'clock we usually go out and finish late chores. Call it that."

A problem of the 1990s is where to sell cream and butter—with eggs, a big source of cash income on the old-time farm. The last local creamery shut down in 1986. "Oh, we're slowly dwindling down," he admitted, putting on a serious face. "Missus likes our own milk and cream. And we've got some chickens. And sometimes a few extra eggs. We've got to help the neighbors out. My wife keeps a big garden and cans. We have a freezer too. TV. The folks didn't have nothing but a washing machine, I guess. An ice chest. I've still got it out there."

I said I noticed he also kept a herd of beef cows. "Gives you somethin' to do the year around, that's for sure," he said. "You don't get lazy, farming that way. If you gotta work all winter. Livestock is like insurance. Crop goes out and you got the cows. You can fall back on them."

Zuckerman says he keeps up on new farming techniques through subscriptions to *Farmer, Farm Journal,* and *Successful Farmer.* "Get them in the mail. You get different articles in there. Don't get to try too many of their schemes. Most of them are for bigger operators. A fellow can read about 'em, I guess."

Sometimes he keeps feeder pigs and usually has about twenty-five sheep for income from lambs and wool. I was glad to hear he opposes confining farm animals. "They confine pigs too much now. Chickens too. Keep 'em cooped up and it's like you get . . . cannibalism, I think you call it. There's nothing else to do. They just sit there in a little space, so big. What are they gonna do? They start chewin' on each other. Some of those birds, boy, they taste blood and they go really wild."

Zuckerman says his 480 acres are enough. "I don't intend to get bigger. My son is getting ready to take over my farmstead. After he gets a little more money saved up, I might sell some land to him. Then, I mean." How did he feel about farms getting bigger and fewer? "Well, it sure would help if some of the farms got smaller. Get a few more farmers out here to farm. Instead of one or two guys farming half a township. It'd sure help the economy in the towns. I'm sure it would."

What matters most about such traditionalism is that it is a form of reaction, of disillusionment with the process of agricultural modernization, the new high-tech farming—call it what you will—and the fear that older values are about to be irrevocably lost in the stampede toward ever-higher yields, ever-bigger machines. I sympathize with the Zuckermans, but there is always the danger with any social reaction that it will go too far, become dysfunctional.

For instance, I defend the family as our fundamental social institution. But it should not become exclusive, swallowing up all other forms of social life. In Prairie, several people told me about Alfred Zuckerman's son, Lowell, the valedictorian of Prairie's 1981 senior class. Janet Schumacher, the wife of one of the more successful modern farmers, mentioned it; her son had gone to school with the Zuckerman boy. "He was almost a genius," she said. "Both of the kids. One girl and one boy. His parents convinced Lowell not even to borrow money and go to college. And they were top of their class. They're both on the farm, sad to say. They've just regressed, you know. They're so old-fashioned looking. And they have such keen minds. My son says the only place he sees Lowell is at auction sales. Or now and again in town."

I asked Mrs. Schumacher what would happen if the Zuckerman children married. "Well, if you never get off the farm, you won't meet anybody," she said. "Oh, they're nice people, the Zuckermans. Very nice. And she'll bring jars of cream to church every Sunday. And anybody who wants a quart of cream can buy it. They have some old tractors to do the field work. Nothing modern." She said the family went to the Baptist church in a little outlying farm community called Cathay. "I hear they go to any activity around Cathay. Bridal showers and weddings and things. I mean they're not exactly hermits. But nobody else farms like that."

Charlie Pritchett's son Jack, another of the Zuckerman boy's classmates, told me, "His folks wouldn't let Lowell go to college or nothin'. It was kind of sad 'cause he sure could have made something of himself. I guess it's just the lifestyle and how he was raised. At the county fair they just come as a family, walk around the booths together, and then all leave in a bunch and get in their car and go home. They're not so—I don't know what you'd say—up to the times. . . . Yeah, Lowell, he just went straight from being valedictorian out to the farm forever."

I didn't go inside the Zuckermans' house. But sometimes you did visit isolated farm families where time had stood still and the sense of stagnation and futility was almost palpable. Once, on a hot summer day, way up in the 90s, I went out to interview an elderly farm couple north of Prairie. We sat around a big square oilcloth-covered kitchen table in a small old-fashioned room just big enough to get our chairs around the table. It was stuffy; no air seemed to stir in the house; and every time there was a pause in the conversation, you could hear a clock loudly ticking.

Despite the heat the old man, aged and bald headed—he sat stiffly erect and hardly spoke—wore long woolen underwear under faded bibbed overalls and an old work shirt which dangled from his worn shoulders as on a clothes peg. His wife, a thickish figure with a wrinkled brow and coal black eyes with pouches beneath them, had drawn her black-gray hair tightly back and twisted it into a roll so that her harsh, grim face was unadorned.

"We've been married 60 years and we lived here that long. And I was born and raised in this house. My parents homesteaded here. In 1893." She spoke with the pride of a survivor. She paused. The clock ticked. "One quarter. That's all we could git. That ain't enough. You've got to have more than that to make a living. Now it's two quarters." She gave a deep sigh, which I took to mean life was hard. Two quarters was 320 acres, also not enough in North Dakota. There must have been difficult years. Like the Zuckermans the couple had kept cows, pigs, and chickens until they got too old and infirm to look after them. The house looked like it was furnished in the 1890s too. Four small square rooms were filled with such relics as an ancient treadle sewing machine, a foot-pumped piano organ, a four-poster iron bed, and faded lithographs of Old Testament scenes in gilt frames. In what she called the parlor was a large grandfather clock. Its pendulum swung back and forth, ticking off the seconds, the minutes, the lives. It filled every silence. "That clock has never stopped running in all my 83 years," the old woman told us with grim satisfaction.

BUD PETRESCU, AN ex-Marine who did recon in Vietnam's Ashau Valley and Khe Sanh, two of the most hellish battlegrounds of the war, fits the 1990s. Now 41, sunburnt, and evidently very strong and healthy, he looks much younger as he takes off his baseball-style seed cap to mop his perspiring forehead. A descendant of one of Prairie's few Romanian settlers, with an open, honest face framed by curly brown hair, he works 3,500 acres, the most of any single farmer. An older brother took over his father's farm near Harvey. So Bud Petrescu decided to be a tenant. It makes him Prairie's prime example of how those without capital can enter farming by renting.

Hard working and always on the run, he takes time out to talk one lunchtime, sitting on the open deck attached to the trailer where he lives right out in the fields with his wife Judy, a pretty, blooming blonde in her 30s who does the work of an extra hand, running the combine or a big tractor. Even the Petrescu's oldest boy, 12, drives a tractor. Everybody has to pitch in.

Over a can of beer, just in from the fields, his faded Marine fatigue shirt streaked dark and light olive drab from sweat, Petrescu talks about how he started farming.

"The way I got into it," he begins, "an uncle of mine had 2,500 acres, an established farm. He was ready to retire, no children. And he tried to rent his land out to his neighbors. But nobody would take that large a unit. He had fifteen quarters of land. And the neighbors would say, 'Well, I'll take these three' or 'I want that four.' And he had a full line of machinery. What's he going to do with that? Have an auction sale or what? And when I got out of the Marines, he said, 'I'll rent you my whole unit, if you buy my machinery on time.' That's how I got started."

In 1988, when the chance came along—another uncle was retiring—Petruscu rented another 1,000 acres, hoping, by farming more land, he could get ahead enough to buy new machinery. He sees little prospect of buying land. "Maybe our farm economy is going to turn around," he said, "but this crisis is going to be with us for many years. I think when the older generation is gone, there's no way my generation's ever going to continue on farming like they did.

"Say this unit I have here. There's no way that I'm ever going to own this farm, you know, be able to buy it and pay for it and continue it. Inherit is about the only way. The last few years, since 1981, our prices have been really poor. And I'm heavily leveraged in debt. I'm paying my cash rent, I'm paying my operating loan off. I'm getting a living off it. But as far as the future goes, on a long-term basis, it really scares me. I'm not replacing any equipment. It's so high priced. There's no way I can go out and trade in a $100,000 four-wheel-drive tractor and buy a new one. There's no way. I've got to keep the old one going and hope that a few years down the road things will turn around. You find every young farmer nowadays is running his equipment right down to the bare end."

The stress on this generation is a big contrast, as Charlie said, to the relative ease of people who got out of farming before the boom ended in the 1980s. "I bet you three-fourths of the land around Prairie is owned by retired people getting their cash rent," Petrescu says. "Sitting around, contented. And we have farmers my age that are farming it. Like I say, the landowners are really well off and the renters aren't. They've got their home, they've got their investments, plus they've got their land. And they're not about to give you a break on this cash rent."

The cash rent most common in the area Petrescu farms north of town is $32 to $35 per acre. The amount varies. Much depends on a farm's wheat base. This is a permanent acreage allotment for wheat and feed grain set by the government to tie farm subsidies to limits on the acreage a farmer can plant. The year 1981 was used. Farmers who happened to be growing beans or sunflowers or other row crops and not much wheat got a low wheat base, which translates into low crop-support payments and a lower rent. Such acreage controls, or trying to limit surpluses by limiting the number of acres a farmer can plant, are controversial. Many argue the government instead ought to limit the number of bushels of grain a farmer can sell at supported prices.

Most renters like Petrescu would like to sharecrop, either one-third–two-thirds or fifty-fifty as mentioned. But most landlords, he says, want a cash rent. "Because agriculture is so shaky right now. If I'm on a share-rent, say two to one, I could get hailed out and the landlord's going to lose his third of the crop also. Or if I'm financially hurting and I'm not able to go out there and apply fertilizer and chemicals, it means less income for them too. They're not going to get their high yield. They know what they can expect in a good year, and they don't want to risk losing it."

Big money is involved. At $32 an acre, payment to Petrescu's uncle with 2,500 acres would be $80,000 a year. Most landlords take half the cash rent in the spring, half at harvest in the fall. But after the 1988 drought, Petrescu's uncle wanted the whole payment up front in 1989. So, as I've mentioned, did one father-in-law from his daughter's husband; the man then went bankrupt.

Petrescu is philosophical about it. "The landlords are afraid if I go broke, they won't get their other half payment in the fall. They're really looking after their own butt."

Petrescu weathered the drought, mainly because, like most Prairie farmers, he had plenty of grain in his bins. "I marketed that grain during the summer of 1988. That July, at the height of the drought, there was a tremendous market. Durum was up to $6.80 a bushel, twice what it was in 1989, a year later." Petrescu says, again like many farmers, that with higher government payments and his grain sales, he actually did better in 1988 when the rainfall was worse. "In 1989 there was no carryover, nothing. The bins were empty. The disaster relief was less. The drought wasn't quite as severe but still bad. So financially 1989 was a worse year."

Among his other landlords are two brothers in California who left Prairie for good when they went away to college in the 1940s. (One of them, a football player, lived with us in Fargo his 4 years in school.) Each of the brothers has held onto 160 acres of inherited land. "You can figure it out," Petrescu says. "At $32 an acre, they get about $5,000 a year apiece. After taxes, maybe $4,000. There's no way they're ever going to sell this back in North Dakota, because it's just frosting on the cake."

Cash rents, he says, are a mark of hard times, as absentee landlords don't always believe the crop losses they get told about. "These last few years, this landlord out in California, he don't know if you're putting his third share in the bin like you're supposed to or you're slipping a load over the hill. So they've gone more to this cash rent." The droughts, however, which got a lot of national television coverage, tended to make absentee landlords more sympathetic, Petrescu feels. "Some of them are really pretty white about it. It's the ones that are living out of state, not real close to the land. The ones right here watching, they aren't giving anything up. No matter how bad it gets, they won't budge an inch."

Just twelve of Petrescu's 1966 class of eighty-five became farmers. Three are still at it. The rest, he says, got too deeply in debt and were forced out. "Not necessarily a bankruptcy. They'd just have an auction sale and sell whatever assets they had and try to clear themselves with their lender. Maybe they bought too much machinery, tried to buy land, a crop failed. Usually they like to stay in the community. But after a year or two you see them moving to a city." There are no jobs in Prairie, he says.

"This is what scares me when you look at my age group. There's nothing in the world that I'd rather do than farm. I have a 4-year degree in Ag Economics; that was my major. Oh, if I had to do it over again, I'd take more agronomy, more soils, more mechanics, welding, what I use in everyday life. But I figure if you own 1,000 acres—six quarters or so—and it's paid for, you can make a good living. I don't want more. I would just love to farm the way I'd like to, set my rotations up and do a good job of it." If a farmer is renting, he says, he needs at least 2,000 acres nowadays. "You're taking care of the landlord too."

Petrescu says it is hard for a younger farmer to bid on land, even if he has the credit. "If a place comes up for sale, you don't see a young farmer going in and buying it. It's usually some older established farmer who feels the land's a good investment. Say there's a quarter, 160 acres, and a young guy really wants it. An older man will come in and pay anything because he's got the money. That's kind of where land's been going. I'd love to buy some land, but you're bidding against the retired banker, the retired farmer. They're looking for an investment, a tax write-off. I'm looking for land to make a living."

Once again it is the theme of generational conflict. But some farmers think Petrescu and big renters like him are better off not owning land. During the 1970s and even the 1980s, people who just rented sometimes made more progress when land values inflated and interest rates were too high.

Petrescu says if he went broke and some big outfit hired him, he'd never do the same volume and output he does now. "What I'm saying is that when you're out at five thirty in the morning, you're not going to do that for somebody else. When the weekend comes along you'll say, 'I'm parking this tractor until you really make it worth my while.' Instead of fighting to keep this old machinery going, I'd say, 'Either you put me something in the field that works or I'm not going to run it.' We have farms in this area that, once they got too large and too many hired hands, the management got bad."

Petrescu says he feels things can only get better. I told him about how so many farmers said they just needed 2 good years to get back on their feet. "Two good years will heal up a little," he said, "but I feel if you've lost something, you'll never get it back." It was different in the Depression, he said. Old-timers tell how bad the 1930s were, the dusty winter of such and such a year, or the summer the grasshoppers came and left the land stripped bare, a playground for tumbleweed. But in the Dust Bowl if you lost your crop, you still had a few hogs, a few cows, a few chickens; you survived. "My Dad says there were some years they never harvested at all. And they lived; they kept the farm. Now if a farmer loses his crop, he's done. He's broke. He's off the land."

HAWK COUNTY LOST about 85 of its 800 farmers in the 1980s. Not all were forced out. Some retired. Others quit when they had something left. In Alfred Zuckerman we met an old-fashioned farmer resistant to change.

And in Bud Petrescu, somebody who exemplifies a younger man trying, without much capital, to break into farming by taking on more and more rented land for his big machinery.

Now I'd like to turn to the successful farmer, somebody with enough land and it and the machinery paid for and with cash in the bank. When I asked people in Prairie who some good farmers were, I usually got three names: Nels Peterson, Paul Fischer, and Harry Schumacher. All three are third-generation descendants of pioneer immigrant settlers in Hawk County, one Norwegian and two German. Each owned his land before the inflated land prices and high interest of the 1970s. Each has expanded his farm sufficiently—Peterson and Fischer to some 2,000 acres apiece, Schumacher to 2,500—to bring sons in with them.

All three do something besides grow wheat. Peterson raises beef cattle. Fischer was the first farmer in Hawk County to raise pinto beans, the kind Mexicans put on tortillas; now he has a seed plant. Schumacher's land is not as good, but he has a large turkey farm on the side and some good rental arrangements. All are prospering.

While I talked to them separately, and each several times, I've grouped them together because their successes have a lot in common. Though all are in their late 50s or early 60s, all church-going Lutherans, and all university graduates, their looks and personalities differ. Nels Peterson, a tall, spare Scandinavian with receding blond hair and a slight sprinkling of gray on his temples, is garrulous and good natured. Paul Fischer is a medium-sized, well-built, brown-haired man with an alert mind and man-of-business mannerisms. Harry Schumacher is big, robust, and beefy. All three are thoughtful, genial men who like talking about farming and ideas.

Peterson lives with his wife, Margaret, and two teenage daughters— three grown sons are married—in one of the few truly charming farm-houses around Prairie, a two-story white frame house faced with open porches and approached up a curved drive through old oak and cotton-wood trees. The Petersons asked me to stay for lunch on one visit, and everybody, plus a bearded fair-haired hired man who moonlights as a bartender, sat around a long table in the kitchen, like something you might find on a traditional farm in Norway.

"You know where we stand now?" half-joked Peterson. "Grandfather is living on $300,000 in CDs. The son bought his land from him at an inflated price. Grandfather went to Arizona, collecting his 10 percent interest. How can anybody be more successful than he? Now his son can't make his payments to the Federal Land Bank. The grandson is now 25.

He's waiting for Father to go broke so he can buy the farm like Grand-father bought it, for $50 an acre." Peterson burst into a deep hoarse laugh. "I've got a son that works in Grand Forks. He says good management is wonderful, but timing is everything."

Fischer, more serious minded, says a farmer should do well on 1,000 acres as long as he didn't buy inflated land or borrow a lot of money or bid too much on cash rent. All three of these successful farmers stress the need for moderation. Schumacher argues a medium-sized farm does best. "It's not just the big rich farmer who has the best performance," he told me. "But a farmer has to have so much land or he won't be around. I'd say a 1,200-acre farm, if the farmer really knows what he is doing, can make it around Prairie. But the key is: how does he get to own 1,200 acres? It takes a lot of time to buy land. He's probably going to rent. But a lot of guys who are cash-renting or sharing are going broke. There's $x$ number of dollars to be had from an acre. I'd say, $30 should be the maximum rent and $25 is realistic. But it's $32 to $35 a lot of places around here. The owner is going to have to get less if somebody like Bud Petrescu is going to survive. I feel that's true."

I asked all three why some farmers wanted to get so big. Were they greedy?

"Greedy? No question it's greedy," Peterson says with a grin. "For a while here the big farmers were really taking over. A quarter of good land would come up for sale. And the real big farmers would outbid everybody else. So they kept expanding and expanding. Leveraging. It's like sitting at a crap table and you win the flop and let it ride."

Fischer is more cautious. "I have no idea why some farmers got so big. I looked at people and said if they had a third of what they had they could still make a very decent living. Sometimes their ego gets involved too. Somebody would say, 'I want to farm a township, that's my goal.' Or 'I want to farm $x$ number of acres, that's my goal.' Most of these people, aggressive, overleveraged, got into trouble in the 1980s."

All three started out much smaller. Schumacher, the poorest when he began farming in 1964, had just 160 acres. Now he farms 2,800 acres, some of it rented, as mentioned. Turkeys were his big source of income at first; he has 13,000 now but has turned his turkey farm over to his two sons, George and Kurt, and George's wife, Sarah.

Paul Fischer used to be a wheat farmer only. He got the idea of raising pinto beans from a friend. "We always met at conferences, and they were growing pinto beans in the Red River Valley and selling them for good money. My friend said, 'You ought to get into them.' In 1974 I decided to borrow a planter from a neighbor. It was a bad year but we learned some things. After that, it just kinda grew. You know, neighbors look across the road to see what you're doing. In time they put two and two together and say, 'He's making a dollar over there. Maybe I can too.' So now there's a lot of beans grown around here. Some people have as much as 800 acres. And there is no other place to market them." By growing certified seed, selling it, and marketing the local bean crop, Fischer's plant serves farmers in several surrounding counties.

Today he keeps up with the Chicago grain market on a computerized service, has a fax machine, and daily reads *The Wall Street Journal*. Fischer provides a good example of how individual farmers in the Middle West, once far removed from foreign events, now find themselves drawn into the new single-world economy. "You have to keep on your toes," Fischer says. "Like in 1981 the Mexican government offered the United States contracts for bean production. They wanted a flood of beans, and our production jumped overnight."

Schumacher, like many farmers, hedges his farm's futures with commodity trading. "Not that much," he explains. "But we've sold some ahead." Until 1972 he had not used the futures market. That year he had 3 years of stored grain in his bins. It was under loan, and people kept telling him, "Prices aren't going to go up. You might as well dump that grain." Schumacher paid no heed and kept building granaries, and in 1973 prices doubled and he made a killing. "It was partly luck," he admits. "But it also cost us money to put up grain storage. All farmers calculate the right moment to sell. But a lot of people don't have the storage, and they're forced to sell whether they want to or not. The banker kind of decides for them."

These three successful farmers are divided on the probable impact of biotechnology. "It's moving down on us like an avalanche," says Schumacher. "It's going to take a tremendous amount of capital. My guess is that unless farming becomes more profitable, biotechnology will not be adopted as readily as it's available." This has always been true, both of science and technology. For example, as Dr. Borlaug mentioned, Gregor Mendel set down the basic tenets of genetics, which led to modern plant and animal breeding in 1866, but his work was ignored until 1900. Hybrid corn did not sweep what we now call the Corn Belt, with its deep, fertile, well-drained soils and hot, humid summers, until the mid-1930s. Dr. Borlaug's Green Revolution, which involves genetically improved wheat and rice seed and more efficient use of water and fertilizer, really the first big application of American farm science to tropical agriculture, did not really get going until the 1970s. It was based on a purely scientific phenomenon: a breakthrough in plant genetics to breed and grow artificially short, stiff-stemmed grain, which can take large amounts of nitrogen fertilizer and water and support heavy grainheads without falling over. I first saw the new dwarf wheat grown on a Delhi experimental farm in 1964; it had first been planted in 1962. By 1969 on a visit to Rome, I found economists at the U.N.'s Food and Agricultural Organization were raving about the new seeds. In the winter of 1970–71, I lived in a Sikh Punjabi village, Ghungrali, in northern India, where the dwarf wheat was being grown on all its 1,400 acres of wheat land for the third year.

This gave one a very good picture of how a new technology gets applied. Schumacher worries only the really big operators will be able to afford to apply new breakthroughs in biotechnology. This is what happened in the Punjab. In Ghungrali, Basant Singh, the most enterprising farmer and, with 32 acres, twice as rich as the average 15-acre Sikh farmer, was the first villager to use chemical fertilizer, in 1960; the first to buy a tractor, in 1962; and the first to buy a combine, in 1970. He sold his first few seed crops of Dr. Borlaug's Mexican-bred wheat at enormous prices, a good example of the rich getting richer.

On my last visit, Basant Singh, who by now had 52 acres, told me, "Even 100 acres is too small. To really mechanize in the Punjab, you should go to 1,000 acres. And we've got to mechanize. It's the only solution to the labor problem." Sound familiar? The world over, the richest farmers are always Basant Singhs. Although vitally important when it comes to innovation and introducing new technology, they have to be kept in check. If he really got his way and farmed 1,000 of Ghungrali's 1,400 acres, it would leave most of its 65 high-caste landowning families and low-caste untouchable and landless Harijan families without a livelihood. Multiply Ghungrali by India's 576,000 other villages, and you can see India cannot possibly follow the American model of ever-bigger and fewer farms. Economically and culturally, its cities would explode.

But you do need your innovators. Schumacher cites banding as another example. This practice, to reduce the amount of nitrogen fertilizer needed in any field by laying it down in strips, is cheaper and reduces the environmental hazard. "The concept came up years ago," says Schumacher, "but farmers weren't willing to buy it when it first came out. You've always got your Alfred Zuckermans, resisting change. Oh, there's some that are going to adapt. Younger ones. It's going to be money if they get held back."

Nels Peterson says virtually all fertilizer except anhydrous ammonia is banded around Prairie. "Whether you're in a 6-inch spacing on your drill or you're on your planter, you're banding." Farmers could do without so much chemical fertilizer, as they did before 1920. But if old-fashioned Alfred Zuckerman can grow so many bushels of wheat per acre by traditional methods, a modern farmer like Nels Peterson can push his yield on similar land up 50 percent by pouring on the chemical fertilizer, by planting his wheat closer together—so close he can't cultivate by machine anymore but has to apply herbicides to keep down weeds—and by using ` pesticides to kill bugs that thrive in congested fields.

"If Alfred Zuckerman puts on sweet clover and plows it down," Peterson says, "he can get by without nitrogen. Look at the size of his operation. He still sells milk and cream and eggs. My dad did that and his dad before him did that. But it was based on cheap labor."

Paul Fischer is the most cautious of the three about biotechnology. "I think the jury is still out," he told me. "Maybe that's just age talking. But I keep hearing people say we shouldn't put all our eggs in the biotechnology basket. I still feel the old-fashioned ways of plant breeding and farming are going to produce many benefits." Which are pretty much Dr. Borlaug's sentiments too.

WE CAN'T FORETELL the future. My guess is another generation of Petersons, Fischers, and Schumachers will be farming in Prairie 30 years from now. But before going into the father-son succession issue, let us look at two farmers who are in trouble. I've mentioned that in contrast to the mainly small-scale farmers going under in the 1950s and 1960s, many of them tenants or regarded in some way as marginal to modern farming, this hasn't been the case since 1980. Many of the farmers who fail now are well-educated, technically efficient younger farmers, with farms presumed to be big enough. So what goes wrong?

For John Farrow, 44, buying a 320-acre farm in 1979 at an inflated $600 an acre—it fell to half that in 1987—was unlucky timing. But the future looked bright and Farrow figured he had little choice. He had farmed an uncle's land for 10 years with the understanding he could buy the place some day. But when that day came, the two couldn't agree on terms and the uncle rented his land to another young farmer. After buying the new farm, Farrow found himself caught in the familiar squeeze, as interest rates soared and farm prices slumped in the 1980s. Worse, he also built a farmhouse, a split-level, yellow-painted showplace that looks like something out of *Better Homes and Gardens*.

Deeply in debt, Farrow survives by working on the side; he sells haying equipment and also serves as a seed company's area supervisor, so in a sense he just farms now part-time. His wife, Anne, went back to teaching school. Their two children are already teenagers.

"I wasn't raised on a farm," says Farrow, a lean, earnest, dark-haired man who, uncharacteristically for a farmer, jogs 4 or 5 miles daily to keep in shape. Anne's grandfather was one of the Red River Valley's big Bonanza farmers. "Those Bonanza farms were too big to last," Farrow says. He got into farming haying for his uncle one summer and just kept working for him all through high school and college. Then, when it was time for him to take over and the uncle asked for more than Farrow was willing to pay, it was over. "Sometimes it happens," says Farrow, philosophical now. "I mean, after it happened to us, we talked to a lot of people and found out it happened quite often. That when it comes down to the nitty gritty and you go to take over, whether it's uncle and nephew, or father and son, you can have trouble. I've heard of fathers who wanted more money than the son could pay. So they split up."

This potential for generational rift is universal. Old Sahdu Singh, the elderly father of Charan, the Sikh farmer in the Punjab I wrote about, was forever interfering. He was a stout, grumpy, remarkably lazy old man with a snowy white beard who lay about on a string cot much of the day, gossiping and drinking endless glasses of milky, sweet tea with his elderly cronies. Charan was quite a drinker when I first knew him, and one time he hid a drum of sugarcane residue in his toolshed, planning to distill some illicit liquor from it. Old Sadhu Singh discovered it and poured it out, declaring, "It does not behoove us to make drink. It is beneath our dignity." Charan, once his father was out of hearing, cursed him violently. "It would have made at least thirty bottles!" he complained. So I was amused to go back 10 years later and see history repeat itself. One day, like Sadhu before him, Charan found a drum of crushed sugar residue hidden in the toolshed and poured it out. These days a virtual teetotaler, Charan told me self-righteously, "I will never allow any son of mine to make this stuff. There are better things to do than waste time making country liquor."

Indeed, generational clashes seem more the rule than the exception. Husen, a poor Javanese rice-growing peasant whose village of Pilangsari I've visited eight times since 1967, had a hard time getting his father, sort of an Indonesian version of Alfred Zuckerman, to plant high-yield dwarf rice and use nitrogen fertilizer for the first time in 1973, tripling the previous yield. But with all their disagreements, Husen showed great respect for the father. We just assumed Husen would take over the father's land some day.

In fact, when the old man died in 1985, there was a falling-out over the land. Husen got his tenth share, but his mother chose not to let him farm the rest but got a sharecropper who was not a family member. Furious, Husen went off to live in his wife's village. There for several years he survived as a hired laborer until, with my help, he bought an unirrigated acre of land. It was a story not unlike Farrow's.

Expressing what is also a universal view, though putting it in American terms, Farrow says, "Farming just gets in your blood." When I asked him why, he said, "You're a jack of all trades. You grow crops. Do a lot of welding. A lot of mechanical work. And a lot of book work, financial decisions. You're your own boss. There are busy times and slack times. Spring planting and harvest, that's your busy times."

Land values, which peaked at $700 an acre in Prairie in 1982, Farrow says, were running about $300, $350, $400 an acre in 1990. "Rent is close to $30 around here. If crop prices stay what they are, I think rents will have to come down a little yet."

Normal wheat yields in Hawk County are 35 to 40 bushels an acre, but they dropped to 10 or 15 bushels an acre in the 1988 and 1989 drought years. Farrow now farms 1,600 acres, renting all but 480 acres of it. He figures the 1988 drought cost him 70 percent of his wheat crop. Like almost everybody else, thanks to crop insurance, disaster relief payments, and high prices for some storage grain he sold, his actual income loss was kept to about 20 percent. In 1989, while he estimates just 60 percent of his wheat crop was lost, less relief and lower prices meant about a 40 percent cut in income. This is about the best statistical summing up on the 2-year drought effect I got. "The drought will have some effect for years to come," Farrow expects.

I asked him if the farm crisis of the 1980s was over. "I do think we've hit bottom," he said. "But it's going to be a slow turnaround. Some farmers in trouble see the handwriting on the wall and liquidate. Before they've lost everything. They can still maybe break out even and not lose too much. Or like me, at some point you look at the whole picture and decide you're going to have to do something. And your wife gets a job in town and you get a sideline. Some guys sell land to pay off debts. You find many renting more and owning less. Some have turned land back to, say, the Federal Land Bank and then rent it back from them."

Farrow remembers how it was in the late 1970s, when land prices were going up $50 to $100 per acre every year. "I mean, bankers were coming out and saying, 'Hey, you want to borrow money?' I mean, they'd just give you money. No questions asked. You didn't have to have a financial statement. You didn't have to have anything. Banks would just give you money hand over fist. You're not going to see that again. Not for a long time. Because farmers have gotten stomped. They've lost too much money. It's going to be slow. Well, it's always gone in cycles and always will. There'll be your good times and your bad times. And hopefully most of us will be around when the good times come back again."

When Andy Rynning, the other young farmer in trouble, got into farming in the early 1980s, it looked like the best of times. Like Farrow, he found bankers eager to loan money. "I mean, you could walk into a bank and say, 'Hey, I want to buy some land.' And they'd say, 'Okay, let's see your net worth statement. Oh, you've got $100,000. That looks fine. You're a good guy. I think we can get along with it.'"

Telling about it now, Rynning shakes his head ruefully. "That was about it. I bought two quarters of land for $800 an acre each. You were allowed 5 percent interest if you were just starting out. Officially it was the 'Beginning Farmer Limited Resource Program' to get more young people to farm. So I got financing through the FmHA. They said, 'No problem.' If they'd really looked carefully at my balance sheet, they would have said, 'I wouldn't loan you a dollar.' But prices were going up. There was the old saying, 'They don't make land anymore.'"

I first met Andy Rynning and his wife, Ingrid, at the county fair. They were a singing act at the grandstand in between horse races. Andy played the guitar and they both sang. One of their numbers was "Tumbleweed." The Rynnings look as blond and Scandinavian as they can be. Her hair is corn yellow, his slightly paler. He is a tall, broad-shouldered man in his 30s, evidently very strong and healthy, and she has a rosy prettiness. They both have the same light-blue eyes. There are two tow-headed little boys with the same looks too.

That day, after they sang, the Rynnings sat next to me in the bleachers, we got to talking and, when I told Andy what I was doing, he said, "If you really want to talk to a farmer in trouble, come and talk to me." The Rynnings live in a modest modern house just beside the old family place built by Andy's grandfather. The farmstead is near a crossroads called South Norway, chiefly known for its architecturally fine Lutheran church. The grandfather, also called Andrew Rynning, learned his trade of carpentry in Norway and as an immigrant found work helping to lay the North Pacific Railroad across North Dakota and build a U.S. cavalry post at Fort Totten. When he had saved enough to homestead and farm, the first Andrew supervised the building of the church. The wood was brought from Minnesota, and its steeple soared to a great height. Today, a century after it was built in 1887, you can see it for miles around. It is easy to see why Andy and Ingrid feel they have such deep roots in the South Norway countryside. (The *South* is because there is a Norway proper about 10 miles farther north.)

It has been said timing is everything in farming. For Andy, 1981 was a bad year to buy land. The two quarters he purchased then had doubled in value from $400 to $800 per acre in 2 years. He figured it wasn't too inflated because his 5 percent interest came to $40 an acre. And he was paying $40 an acre rent. Why not, he reasoned, pay it as interest and buy the land?

It didn't work out that way. "The bigger farmers bid up the prices," Andy recalls. "They had the paper assets. We got into a bidding war. And I didn't even get the top bid. The guy that sold me the land gave it to me because I was a young farmer just starting out. I wish he never had."

Things soon went wrong. Andy says he had a bad year in 1982. "Sunflower prices and sunflower yields were down. Mustard got into the fields. Everybody had it. There was no chemical to control it. When it goes, you get a panic situation. Things just plummet, plummet, plummet."

He sold one quarter to a brother-in-law who had bought the local Norway bank. Then the bank itself closed. "There were a lot of bad loans," Andy says. "He financed some big farmers around. A couple I know, he was really bragging them up. 'You've got to be big to be efficient,' he'd say. 'It's the only way.' But bigness and efficiency take a lot of capital, a lot of machinery. They got machinery-itis."

Today Andy farms 1,200 acres; he is trying to pay for 300 acres of it, the rest he rents. Like others, he doesn't mind absentee landlords and says with a grin, "They're easier to get along with." He and Ingrid are active in the Farmers Union. "There's a lot of debate on the old question: what's a family farm? The statement they've had for many years—and people are finally listening—is it's a do-it-yourself farm. You and your family do it all yourselves. Around here that's mainly 1,000 to 1,200 acres."

He figures two-thirds of the land in his area is rented, a third of it owned. "The Farmers Union wants family farms, and they want the ownership of the land to stay here. I'm probably a thorn in their butt sometimes, but I don't think you can do it. The Farmers Union philosophy is too idealistic. The Farm Bureau is too hardheaded. One says, 'Let's look at the morality of the farm issue.' The other says, 'What's the business side?'"

Andy says he lost half his wheat crop both years of the 1988–89 drought. He was sure it would wipe him out, but the FmHA restructured his loans. "If you have the attitude, 'Hey, I'm not going to give up and quit, I'm going to continue to farm,' there's all sorts of ways to work it out. Everybody's in the same boat."

Andy is one of a growing number of farmers to use a computer. He says, "It's the best piece of machinery I have." Computers, to store and retrieve farm records and information, while fairly common around Prairie, have not taken over as much as once expected.

Ingrid Rynning feels they are still not out of the woods. "We nearly lost our farm. We may yet. Things keep changing, policies keep changing. Andy's mom mortgaged some of her land to help us keep ours, and we rent the rest." In Norway the Rynnings belong to a group of young Lutheran and Catholic farm couples who are all going through financially hard times. Ingrid says, "I think a key to surviving on a farm is what kind of support you get. You need to talk about your problems. Not too many have left the land. But there are plenty who are hanging by their fingernails."

She finds attitudes have changed since the late 1970s. "Then it was cutthroat farming. No question. Your neighbors didn't count. You didn't care if you stepped on their toes. I didn't like that. In fact, the women, I think, disliked it more than the men. It was the *Successful Farming* way of farming, I'd say. It was: 'Go out there and get as much land as you can.' And now that everyone has been hurting, they're coming back to the neighbors. No more cutthroat farming. At least, not for them."

The emotional stress on families is the worst aspect of the farm crisis, Ingrid feels. "The worst of it, 4 or 5 years ago, it was . . . hell. Grim. Toward the end of winter can be depressing. December, January. The rents are due. The loan payments are due. Your husband's got to start lining up next year's operating costs. Plus chances are it's very cold. I never experienced anything near poverty. But one time I did have to go into my children's bank to get money to buy milk. To me, that was pretty scary."

"ONE GENERATION PASSETH away, and another generation cometh: but the earth abideth forever." In the time of *Ecclesiastes,* there were plenty of male heirs to take the land over. Today, with nearly city-sized farm families, this is no longer true. More women are farming (in 1980 up to 10 percent nationally, but just the few that work with their husbands in Prairie or Crow Creek). In any case, the decision by so many farmers and their wives to have fewer children is beginning to reshape the father-to-son continuity that in agriculture goes back to biblical times.

Of course farms are bigger and fewer; there is less land to go around too. Look at the two sons of Harry Schumacher. From George Schumacher's class of thirty-six in 1982, six are farming. From Kurt Schumacher's class of 30 in 1985, two are farming.

I asked Harry Schumacher how young people get into farming. "They'd have to start with their parents, I think," he said. "Around here, I'd say, the farm has to be big enough. Then the sons can be worked into it. The first few years they'll have to be subsidized by the father. There's no way they can strike out on their own and buy a big enough farm. As they profit, they buy machinery. My boys, George and Kurt, rent some land from me. And they run the whole turkey operation. Sarah's kind of in charge of it, and she does all the records."

Schumacher says if he were looking for a farm to buy today, he would look at the size of the nearest town and how far away it was from the farm. "If it's over 10 miles, I'd reject that farm right away. It's not just the shopping community. It's schools and churches. Your wife might put 20,000 miles on her car a year. All those extracurricular activities. You have to get them there."

Schumacher thinks that if a son has to go to work in town until his father is ready to retire, he's unlikely to come back. "I'll tell you what happens here," he said. "If you leave the farm for 2 or 3 years, it's almost a whole new learning experience when you come back. You come back and everything has changed. Chemicals. Ways of doing things. Like what's the seeding rate of wheat per acre? Or how much fertilizer should I put on this because last year we grew that on it?"

But sometimes, he went on, a farm isn't big enough when a son reaches the age to start taking over. The son gets married. Suddenly there's more than one family. "The farm is too small," Schumacher says. "Then everybody goes broke. The farmer has got to increase the size of his farm *before* the son moves in on it. And I don't see a lot of farmers doing it. Maybe the father's in his 50s and he's reluctant to expand at that age. And that's when the son wants to move in."

A perfect example of this is the Whites, father and son. Jack White has about a hundred head of beef cattle and 750 tillable acres, plus some pastureland, south of Prairie not too far from the Schumacher farm. His son, Chuck, rented another 250 acres of his own, but the only piece of land he could get was 8 miles away.

Chuck White, a curly-headed youth, not tall, with an open expression, told me his story: "I wanted to farm. I couldn't farm just working for wages for Dad. And I couldn't make enough just on cows. Where do I turn? Do I get a job in town? The elevator's expanded. They've got thirty or forty people in there. They bring the wheat in on semis from substations in smaller towns like Heimdal and Hamberg and Manfred and Bremen. Then Prairie ships it out on twenty-six or fifty-two-car-unit trains. It's pretty much just a change from a bunch of independent elevators to one big co-op owning it. Right now I could get maybe $12,000, $14,000 a year working at the elevator. It's moderate; it's not terrible."

Instead Chuck decided to grow cantaloupe. It was a gamble. North Dakota averages 110 frost-free days. Melons need 90. "So it might make it and it might not," Chuck said. "You keep asking: when's our fall frost? When's our spring frost? How late can you take 'em out? How early can you put 'em in? I didn't want to rent a whole lot more land. Melons had the potential on just 2 or 3 acres to earn what 100 acres of wheat can."

The first 2 years he got back $5 on every $1 invested, and in just 4 months' time. He decided, along with the melons, to rent two quarters, or 320 acres, of land. Some of it was $30, some $35 an acre, too high for its productive value, but competition was stiff. Plenty of farmers, especially the younger ones trying to afford to buy new machinery, want to rent more land. "The nearest I could rent was 8 miles from home. It was hard dragging a machine that far. We weren't set up for hauling back and forth. People get desperate for land around here. Their banker is demanding payment. They need money, cash flow. They've got so much machinery. Why not go from eight quarters to ten? Eleven? Hock up a little more land? Well, the next man, his banker is telling him the same thing. So they're all out there, competing for land."

Chuck got married 5 years ago. When their first baby came, he and his wife took stock of the future and decided he ought to go back to school. As Chuck says, "To get that 4-year degree, that sheet of paper that will enable me to go out and get a job."

Jack White, his father, still hardy at 56, told me, "If I could, I'd figure out a deal for Chuck to take over so I could quit working 12-hour days. But the economics just aren't there." He tells himself Chuck can only benefit from more education. "Maybe we'll still be able to figure out a way to make it work. But for now, I've got a debt and it's got to be paid." White feels aggrieved that after the drought some families got their farms refinanced and reappraised and the fathers got out and the sons got in. "Do you feel that's fair, really?" he asks. "To me, it's not fair. I gotta pay my debt."

There are a lot of Chuck Whites, going off to a university reluctantly, learning an urban profession, and never coming back. Some young men, like Charlie Pritchett's son, Jack, take on several jobs to enable them to keep living on the land. Jack, a tall, broad-shouldered, fair-haired youth with regular features and his family's good looks, is a crop appraiser for a private insurance underwriter, does taxidermy on the side, and helps out at Prairie's fertilizer depot in the fall and spring. Charlie rents his tillable land, but Jack and his wife and baby live in a trailer out on the old family homestead, using some pasture and a couple of the barns. Jack and his wife, a college graduate who works in Prairie's bank—Jack himself went to college 2 years—are building up their flock of two hundred sheep and also raise ducks, geese, and some Arabian horses.

"I guess I'll keep expanding," says Jack. "Get more sheep. This is making money." He says that while a few of Prairie's young men never want to see a tractor again, there are many more who would farm if they could swing it.

Knute Hanson, lanky and calm mannered, with a thatch of black hair, the son of a retired county agent, feels the same way. But, he says, you really have to know what you are getting into. In his 1978 class of forty-three, three are farming.

"Young people, if they want to farm, they have to make a great commitment," he told me. "I rent my place. Grow wheat, corn, sunflowers, dry edible beans, and barley. We have a 50-acre corn base. I was growing sunflowers in 1981 and got a low wheat base. You get a good wheat base and a farm would probably be worth $40 an acre to rent it. I've got a cow-calf operation, about a hundred cattle. It's a safety net if you have livestock. You can chop and salvage the corn in a bad year. You can get grain, get a loan on it, buy it back, feed it to your livestock, and still get your payment. Like this corn. I could chop up all this corn and feed it to the cows. But I'd still get paid the support price."

If rents were really fair around Prairie, Hanson says, a farm with a low wheat base would go for as low as $25 an acre. He and his neighbors pay $30. "My wife teaches school. That's where our living comes from. Everything I make from the farm goes back into the farm. I don't own any land. And my tractor has 10,000 miles on it. It breaks down a lot. It's broke now. If you've got the money, and sell your equipment before it depreciates too much, you might save by buying new. It works for some people. More common is getting older equipment you have to repair. I've got everything I want."

Prairie's high rents are one sign a lot of young men are trying to get into farming, or farm more, like Bud Petrescu going from 2,500 acres to 3,500 to try and make enough to buy new equipment. "If I plan on having kids," Knute Hanson says, "I've got to have something big enough going so I can have the kids work on it with me."

An older Hanson brother, 32, farms for a grandmother 45 miles north of Prairie. A third Hanson son, Mike, just in his first year at the state university in Fargo, wants to farm too. Very young and tall and solidly built, with a sunburnt, childish face and a trace of chin whiskers he will soon need to shave, Mike feels, as I do, that the generosity of the parents matters a good deal. "To go into farming, you need help from somebody like my dad," says Mike. "Somebody who cares. Either you gotta be a real go-getter or you've got to have somebody to help you out, get land and machinery and work a partnership of some sort. 'Cause otherwise it's pretty much impossible." In the Hanson family, their retired father handles marketing and managing the government programs. Mike says for one of the sons to do it would take half their time.

Mike also feels anybody who wants to farm nowadays is wise to get a college degree; the vast majority of Prairie's farmers have them. "A high school dropout would have a hard time getting anybody to rent them land. They have to be willing to take a chance on you." Mike is another of Prairie's young men who feels a farmer doesn't need to own land as long as he can rent it. "I bought some land when I was a junior in high school," he told me. "It was about the dumbest thing I ever did. I paid $450 per acre for an 80. It's only worth $300 now. And I don't own it. The Federal Land Bank owns it. It's crazy if you start borrowing to own land."

When I asked if too much money went to landowners, Mike Hanson's answer was typical of the attitude of most farmers: "I guess to me what's yours is yours. And it should stay yours. A lot of these old retired people worked very hard for their money. Land is just like money in the bank." And the government's farm subsidy program? "Some it's saving, some it's helping, and for some it's just gravy."

Both Knute and Mike Hanson feel many parents in effect subsidize their sons to farm by giving them loans at low interest to buy land and machinery. "The parents don't tell the neighbors they're giving the sons anything. But that's what it is," says Mike. Knute agrees, but repeats what many have said, that timing is what matters when it comes to starting out. "I guess what I'm hoping is that my timing for getting in is going to be right. It seems like anybody who gets into farming at the right time, they can make it. It's yet to be determined if this *is* the right time."

Mike feels even some farmers with big debts will hang on if they possibly can. "Just changing a job, period, is hard. I think it has a lot to do with pride. And not knowing what to do. Well, look at yourself if you were farming and 40 years old. And you've got a couple kids and that. You don't want to leave. And they'll stay out there and beat their heads against a rock if they think they can hold on."

Every younger generation starts out more idealistic than the one before; I think Prairie's young people see themselves this way too. Jerry Fischer, another sunburnt, tall, broad-shouldered youth with clear, clean-cut features, feels his generation is less driven than his father's. "Back in the 1970s," he says, "farms just kept getting bigger and bigger and bigger. Now young people are saying you don't make that much more money with an extra quarter. Your machinery depreciates so fast, you can't afford to put those extra hours on it."

His father is skeptical. "Bud Petrescu might talk about wanting to farm just 1,000 acres, but he'd have time on his hands," says Paul Fischer. "Go back and see some day. Because it's built into a farmer, if he's got a little bit of time and if that quarter 2 miles down the road becomes available, that farmer's going to think, 'Maybe instead of making *x* number of dollars. . . .' And it's the same old vicious circle."

IN PRAIRIE WE'VE met successful medium-sized farmers, farmers in debt, one who clings to the old ways, another who rents all the land he can, and these youngsters trying to get into farming. Our last farmer is the Basant Singh of the piece. As somebody said of him, "Ernest is always on the cutting edge of whatever's happening."

Midwestern farmers tend to look alike in their seed caps, white foreheads above sun-weathered, windburnt reddish brown faces. The young, mostly of Scandinavian or German descent, are often blond and fair, with muscular builds and open, honest faces. The older farmers are putting on weight, wear glasses, get crow's-feet around the eyes, are losing their hair. The very old, of course, right out of Shakespeare's seventh age, get lean again, gaunt, bent, toothless, their faces wrinkled and creased as walnuts from a lifetime of prairie wind and sun. It is the same with the women. You will never see so many lovely young girls as in North Dakota; it strikes a visitor every time. Fair, with their long blond hair, and radiant with a kind of rosy Nordic prettiness when still young and fresh, they, too, vanish by middle age into plump and hearty farm wives, with glasses and frizzy hair and voices grown hearty and loud.

So it is a surprise to meet Ernest Waldenberg. He looks like Napoleon. Maybe it is his shortish stature. Or the way his coarse curly black hair falls across his forehead in unruly bangs. My guess is it is the glint of crafty intelligence you catch now and then in his eyes. And his air of complete self-assurance. No wonder. For he is Hawk County's biggest landowner with 9,129 acres, about 7,000 of them tillable. Nobody else even comes close. The runners-up, several brothers who farm together, have 6,500 acres between them.

There is talk in North Dakota of farmers owning a whole township. That would be 23,000 acres. Nobody around Hawk County has that. So many of the really giant farms went broke in the 1980s. Nonetheless, Waldenberg's domain, down in the southwest corner of Hawk County, is plenty big. Besides his land it includes an elevator he presently leases, several trucks, a hail and crop insurance agency, and a combined grocery and hardware store that supplies local farmers with fertilizer, pesticide, and seed.

"I've been farming between 6,000 and 10,000 acres for the last 15 years," he told me. "I think we have a very efficient farm. It's multimanaged. My son and I and two full-time employees. So it's a couple of thousand acres per man.

"Our hired men have a better standard of living than if they were farming on their own. I think they have the same commitment we do. They don't want an 8-hour day. If they want an 8-hour day, they can't work for us. Farming, you can't do that in the summer. It's on a year-to-year basis. In salary and bonus and housing, our two full-time men get in the neighborhood of $25,000 a year. The one man we furnish housing; we provide heat, lights, everything but the telephone. We consider that comes to about $200 a month. We also cover hospitalization. Their monthly draw is like $1,200 a month. And we pay them a $2,000 to $5,000 bonus at the end of the year."

In flat wages, it comes to $14,400 a year. Waldenberg says both hired men have been with him 6 or 7 years. One is a cousin's son. "His dad worked for us until he went farming on his own. Didn't succeed. In fact had a lot of financial problems. And then his son came back and asked if he could work for us again. He's really our mechanical man. Machines, electrical—Lloyd's the man. The other fellow is a local. He's about 45. He just came and asked for a job. He'd worked in roof construction and wanted something more stable."

When he was younger and farming was less high tech, Waldenberg says, he just hired high school kids in the summer. "We'd pour on the coal for 3 months, then I'd drop back to one man in winter. But now that we're diversified, we keep those full-time steady men and hire a couple of extras in the summer."

How big can a farm get? Waldenberg says any farm, if it gets too superbig, can become inefficient, take on too much debt, and lose management. He notes that several families in Carrington and Jamestown came out of the 1940s strong and, using their land as collateral, got really big in the 1970s. "They were big and then they went out. Overexpanded."

Around Prairie, he says, you won't find really big farms. "I mean, we put together a lot more land down here than you could do around Prairie. Number one, I think, was the community attitude. If you expanded like I did, you were a bad boy. You go south of Prairie, Elm Township was kind of the Mecca of the Farmers Union. If you got over four quarters, you were a bad guy. That's down where the Schumachers and Whites got farms. I had a friend there, he started out in Elm Township in the mid-fifties. He got so frustrated if he tried to buy a quarter of land. He figured he needed five or six quarters to farm. So he quit farming and went into insurance. He was the one with vision. But he couldn't expand the way he wanted to."

Another example, he says, was somebody like Paul Fischer. "Now, as far as I know, he inherited his land. He's got the same land his family had. A very good manager. But just that little pinto bean operation. He created that seed business. But he didn't add land. He was a very conservative manager. And to me, that fits Paul's personality. He's a meticulous, careful person. Enterprising, but not when it comes to expanding. You see, Paul could've. . . . If his personality had been different, he could've been farming 5,000 or 6,000 acres. And reaping rewards two or three times as much as he does now.

"Of course, that's the route I took. But I didn't try to farm 100,000 acres. One man at Jamestown tried to. But in order to farm 100,000 acres, he had to farm all the way from southwest of Bismarck to north of Grand Forks. He even farmed 2,000 acres east of Prairie. Well, he was the first one into bankruptcy."

Have we seen the last of one man trying to accumulate so much land? "Oh, people will be a lot more careful for a little while," Waldenberg says. "But you forget. The opportunity was there, you see. These people are adventurers. Maybe a person's not satisfied with $1 million worth of property. He wants $5 million." Waldenberg asked if I'd ever been to Medora, the little town in the North Dakota Badlands named after a legendary French noblewoman. With her husband, the Marquis de Mores, she tried to build, while fighting off Indians and outlaws, a nineteenth century cattle-raising empire. I'd driven through the Badlands, a weird and desolate land of grassy plateaus cut by deep coulees. The Medora saga, like the Earl of Selkirk's doomed attempts to set up a farming colony in Dakota Territory, was as romantic as any fiction. "The Marquis de Mores," Waldenberg says, "squandered a million-and-a-half dollars of his father-in-law's money on a packing plant. And in 5 years he was gone. Eventually he died; somebody killed him, you know, in Africa."

Human character, Waldenberg asserts, doesn't change. He sees it as a member of the board of directors of a Harvey bank, he says. "Okay, you get a farmer in trouble. And you devise methods to get him out of trouble. If the economy improves and you let up on him at all, he's going to be right back in trouble. In the same boat." I remember when the landless field workers in Ghungrali struck for a higher wage, Basant Singh told me, "Drink is their problem. That's the touchstone of all their problems. They won't change."

His only son is a farmer, Waldenberg told me, but one daughter is married to an electrical engineer with Texas Instruments and another is a computer scientist out in Pasadena. "The thing I really appreciate about America is our freedom of choice," he says. "I haven't done a lot of traveling, but I've been in Western and Eastern Europe a few times and

Russia once. So I guess I'm one of those 24-hour experts. But I came back from the Soviet Union with such a strong feeling of how fortunate we are that we can try to find in our lives *what we like*. That we aren't locked into a profession because we were a doctor when we were 24 or we were a farmer when we were 18. Maybe we can be a farmer for 10 years. We can be a car dealer for 10 years. We can go broke. Or we can succeed. We can pick up and start all over again."

He finds even Western Europe very traditional. "I couldn't believe it. You go 20 miles and a different language is spoken. I came back feeling all this melting of different traditions is what's really made America the strong power that it is. We adapt more quickly."

Waldenberg believes the farm crisis won't last too many years more, though we may not have seen the worst of it yet. "We're going to adapt. We'll figure out how to get more farm products into industrial uses. We're going to compete with other countries so they stop expanding their production. We're going to quit giving them the advantage."

As long as Minneapolis keeps pulling its money back, Waldenberg says, you know a recovery hasn't started. "Just like it pushed money into North Dakota when agriculture was expanding in the 1970s. When agriculture is retracting, you bring all the money back in. They're selling banks to local people for 50, 60 cents on the dollar. Because they don't want to take the risk. You can build a little Hardee's hamburger joint outside Minneapolis and make more money than on a 5,000-acre farm in North Dakota. So who do you loan your money to?"

Isn't it shortsighted, I asked.

"Sure, it's shortsighted. The Minneapolis bankers should stop and think: 'Services will only carry us so far.' I mean, they'll all end up doing each other's laundry. They're not thinking about that though. Trouble is we can't all do each other's laundry. It's imperative, if we're going to keep up any semblance of our standard of living, to forget about doing each other's laundry."

How bad is the farm crisis?

"If you look at the problems of the farm credit system and farm equipment sales and business failures in small communities, it's hard to exaggerate it. There are small towns dying. But they started dying in the 1930s, and some of them still aren't dead. Take a look at the nearest branch-line railroad. All the little towns along it were started up in the 1890s when the Northern Pacific extended its line from Carrington. Well, go down the line. I guess for all practical purposes the community of Heaton is dead.

The elevators are closed up. The bank is gone. All that's left are four or five houses and a church. No services.

"Now if you come up the road, Bowdon has survived. It goes up and down, has services or not. Chaseley, the next village over here, may be the next to die. Here in town I own the store. I own the elevator, though I have it leased out. I've been hearing for 20 years that we're on our last legs.

"My little store here is a people business. And people businesses go downhill. The acre businesses go uphill if they adapt to the times. What do I mean? The big growth has come in inputs per crop and handling those crops after they're put in. Up until probably 1977, in this area we didn't have any demand for seeds and chemicals. Then row crops like pinto beans and sunflowers came in. And we had a chemical and seed explosion. By the early 1980s we were moving toward continuous-crop farming instead of fallow farming. We were moving toward higher cash-input agriculture. In fallow farming, we'd left a third to a half of the land lay. Kept it black dirt to rebuild the soil fertility. We didn't take as much out of it. It was low cash-input farming."

Waldenberg feels the trend is for more people to get jobs in services like grain elevators, but with fewer people on farms. "Maybe you lose five farmers and put one more man in the elevator. That's a net loss."

Waldenberg says the Farmers Union is afraid big corporations will take over farming. "I say that corporations will never take over farming. Look at Cargill. They don't want to bother raising crops out here. They don't have to do that. It's too unreliable. They can't manage it. No, they want to control it. And that means to control the inputs and output."

Harry Schumacher told me much the same thing, that what farmers bought and sold could be controlled by big corporations. "They'll squeeze the daylights out of the farmer," he said frankly. "And he has no pressure to bear. There's no way he alone can do anything. They'll squeeze him until he's just about busted, and after he's busted they might try to take over his land, if possible, like banks did in the Depression."

The afternoon of the day I went out to see Ernest Waldenberg, I was asked by Martha Adams, a home health aide with the county social services office, if I'd like to ride along with her on her rounds. She looked after a number of old people around Prairie, giving them a bath, doing some of their housework, shopping, sewing and mending, giving them rudimentary nursing care. It was the kind of thing that enabled old people to stay on in their own homes.

It wasn't an easy job, and on the way home I asked Mrs. Adams, a cheerful, gray-haired woman in her mid-60s, how she got into it. Her husband, a lifelong farmer, she explained, went broke and, on top of that, had a heart attack. After selling some land, they were left with a more-than-$100,000 debt. Essentially unskilled, Mrs. Adams made $4.69 an hour and worked between 20 and 40 hours a week. In 12 more years, at age 69, she could expect to make $6.21 an hour or, if she chose, retire on about $250 a month. Still, she managed to pay so much a month on the interest of her husband's debt.

"I loved being a farmer's wife," Mrs. Adams told me as we drove toward Prairie through gently rolling hills, scattered ponds, and fields of black earth and the soft green of quickly sprouting grain. "We never dreamed it would end this way. We worked hard, got all our children through college."

That same evening—it was a full day—I was invited out to men's night at the Prairie Country Club. It was a warm spring night and we sat around with drinks as golfers drifted in. The talk was about some big farmer in Harvey, known for going on safaris in Africa, who, after the 1988–89 droughts, had a large government debt forgiven. When I mentioned I had interviewed Ernest Waldenberg that day, somebody said, "Did you hear he got a $300,000 writeoff?" Everybody was mad about it, though it was all perfectly legal. I mention it not to cast doubt on what Waldenberg said—he is shrewd and sees things from a valuable perspective—but after hearing Mrs. Adams' plight, to leave it out would not be to tell the whole story.

"It happens all the time," one of the country club crowd said. "And not just in Prairie."

IN *OUR TOWN,* first performed in 1938, Mrs. Gibbs mentions to her husband that some people are starting to lock their doors at night. Dr. Gibbs says, "They're getting citified, that's the trouble with them. They haven't got nothing fit to burgle and everybody knows it." Fifty years later some of the people in Prairie still don't lock their doors. An old saw is that even a broken car window is news.

In reality, says Hawk County Sheriff William Jager, a good talker some call Windy Bill, there is a good deal more crime than that. To be sure, deadbolts are unfamiliar, nobody has to keep an eye on her purse, there are no weirdos in the street. That sense of looking over your shoulder at all times, like you get in some cities, is missing. In Prairie crime goes up and down depending on how tough times get to be. In 1987, the worst year of the 1980s, thefts on farms and burglaries of stores and elevators went way up; by 1990 they were down again. The same with bankruptcies, fore-closures, and auction sales.

A growing, relatively new crime is linked to demography, as so many changes are. With better roads, more off-farm work by family members, multiple-tract operations, and fewer farmers keeping livestock, there is a trend to live in town. By 1980, nearly a third of farmers and three-fourths of farm workers no longer lived on farms (something Rev. Williams, for one, never expected to see). As Sheriff Jager says, "It makes it very tempt-ing for crooks when somebody leaves all their tools and equipment un-tended in their sheds and barns."

What turns a Prairie youth into a criminal? Years ago I made a study of the criminal underworld of the Casablanca waterfront. If you were sympa-thetic, you might say a poor Moroccan, maybe with a hungry family, could be desperate enough to try anything. If you were hostile/afraid, you blamed criminality on childhood experience, psychic debility, or even derangement. Or you blamed, as I would at least partly in American cities today, the cultural breakdown, particularly the family breakdown, of the larger society.

In Prairie, as everywhere, there are elements of all three. Sheriff Jager says theft goes up in hard times, there is no question about that, but not necessarily violent crime. "As far as we're talking murder, rape, assault. Those crimes have nothing to do with farming. If I want to murder, I'm going to murder. If I'm a rapist, I'm going to rape whether I got good times or bad times."

The theft of tools or equipment from a farm is likely to be local and economically motivated. "Maybe not in my county, but close. More than likely I'll know them. You just find these guys that are on the streets driving all night. If I had to guess an age, I'd say 26, 27. High school dropouts. You're going to find the guy at the lower end of things. The guy with the minimum wage. He's got no skills. He's got no nothing. And after he's got a record or been in trouble, being in a rural community as we are where everybody knows everything, he's got no job. So Johnny gets arrested for stealing in Harvey. He comes down here and tries to find work, even as a farmhand. And we know he steals, so we don't want him around. So basically the only job he's going to get is a ditch-digging job.

He's broke. He needs money, same as I do, to eat. So what does he do? He steals drugs or gets his family to support him. And once he gets a jail record, that makes it that much tougher for him. And it's just a downhill slide."

Murders Jager sees as the work of a psychopath and unrelated to ordinary crime. "We've had 'em. I'm going on 15 years as sheriff and then 2 years deputy. I've had, I don't know, six, seven rapes, murders. It's been 9 years since we had that double homicide. A guy murdered two Indians, Sioux. Picked them up at Pingree standing in the rain. He wined and dined them through the area. In fact I was in the local tavern when they came in. He brought 'em into the Cabaret and bought 'em drinks and stuff, and they stopped again at Artos on the highway outside Harvey for more drinks. Then he took 'em west of town and shot 'em both. It didn't seem to make much sense. He just stood them on the highway and shot them."

If murders are rare, drunken fights are not uncommon, especially at the county fair. "Ten years ago we had a carnival come for the fair that brought in a whole motorcycle gang. Working on the rides. I mean, they had long chains welded to their body. They were rough. Of course they were in town 5, 6 days. Fights, faithfully, every day. And busted noses. It got to the point, you know, when the townspeople didn't want that carnival hired back. More and more of these carnivals pride themselves on a clean midway. Oh, some years we'll backslide a little. You get 'em where they don't have shirts on in the sun, wear earrings, and look a little rough and weird."

Jager, a heavy-shouldered, powerfully built man in his 40s with receding blond hair and a starched uniform, looks able, with his revolver strapped to his side, to handle himself.

I asked about drugs; we were about as far from the relentless scream of city police sirens as you can get in America, but somehow drugs are seeping down everywhere. "It's on the downswing," Jager said. "Due to the price of it. Good marijuana, if you're talking Colombian, Hawaiian, or some of the fancier cuts, you're looking at $250 an ounce. Where does a 14-year-old get $250 to get an ounce of marijuana? They don't have it. The kids are all going back to alcohol. We've got a few of that refined crowd in their mid-30s that are into cocaine, a little fancier stuff. They're not the long-haired creep in the street. They got money. You never see the drug."

Jager says at a big teenage "keg party," as many as 300 or 400 people might show up, coming a long way from several neighboring counties.

"There you're going to see a mitch-match of everything. The majority's going to be drinking beer. Some on marijuana. You're gonna have the oddball now and then on pills and stuff. You're going to have a few on cocaine. But the drug problem we knew in the 1970s is basically dying."

Then, he says, you even had older, high school kids giving drugs to grade school children. "They were using kids as guinea pigs. To test the drugs. So we started gearing up, telling the kids: don't take pot, don't take pills. We start with the first graders, third graders, and seventh graders. We give 'em classes. And, I don't know, it seems to be working.

"Sure, right now, today, I can name people in town that are dealing drugs. To supply their own habit. We call 'em 'creepy critters.' High school dropouts. Long hair, earrings. You just see 'em on the street. When everybody else goes to bed, those rats start comin' to life. Sleep all day, cruise all night. Come from some of the best of families and some of the worst of families. You can tell by the cars at their houses. You know, they're marijuana users. They got to get a few bucks so they can afford it. Oh, as I say, we had 'em all the way down into the lower grades when pills were a big item. Dealers pushing it in the schools. Tough times have ruled that any high school kid smoking marijuana now, he's growing his own. He's got a little patch here, a patch there. Very poor quality. Nothing to get excited over. You go to the Colombian, the guy that's smoking that is the dealer who can afford it; probably he's selling junk and keeping the good stuff for himself."

Jager feels a town like Prairie is about 10 years behind the national average when it comes to social trends. "If California's got something new, 10 years we'll have it."

When it comes to pornography, the sheriff doubts if you could even buy a copy of *Playboy* in the county. "As far as that, I know guys that's got 'em, got some of the roughest things you can imagine. But as far as going and renting a video. . . . Yeah, for the stag party, the bachelor party, we got one guy in the county that's . . . I mean . . . he's got a collection of filth that's just . . . every rotten . . . that you can dream of. . . . He's just. . . . It's just the way he is. Everybody that's going to have a stag party runs to his house to get these VCR orgy things. Everybody wants to get them for these parties.

"Matter of fact we had a stag party here a couple months ago. I never even heard about it until after it was over with. One guy was having a bachelor party, so they brought a couple strippers from Minneapolis, headed out for the farm, set it up in a quonset. A lot of local guys were there, and of course I heard back from the wives' side that a lot of husbands were in the doghouse. They went to a bachelor party and they figured, well, the most it would be was a dirty movie. Well, it turned out they had a couple ladies running around naked all night long. That didn't set too good with the wives. But that's a rare occasion. I'd say the morality here in general is pretty good.

"Everybody knows everybody. I know I got a couple of brothers in Denver. I asked 'em, who your neighbor is. Who is so-and-so on the side, you know. They said, 'I have no idea.' Didn't know. Didn't care. That don't-care attitude. But *here* if you go and you're a rat, everybody in town knows you're a rat; they treat you like a rat. Here, if you're a farmer and you break your leg today at work, you're going to have ten neighbors coming in to combine the whole crop off, free of charge."

Jager says such ethics go with a farm community. But it works only in Prairie itself. "As far as you go to Arizona and you get into trouble, your neighbor up here ain't going to do much for you. He didn't get to Arizona. He doesn't have that big Plymouth you own. So you're not going to find him coming down to bail you out." In effect, Jager is saying that the agricultural moral code is rooted to a specific rural group. "As far as if you're injured or hurt or you've passed away and your wife has to farm here at home, the neighbors come in, farm the land, do things for you." One also sees this in the Third World. In Cairo, for instance, urban migrants tend to cluster with their fellow villagers from home, and some-thing of their village's mutual support system is carried over into the city. But when a villager goes off by himself and gets into trouble in Cairo, he is pretty much on his own.

One of a county sheriff's jobs is to collect unpaid bills and to foreclose on farms. In 1987, when I first talked to Jager, foreclosures in Hawk County were running about one a month. By 1990 it was less than half of that. It is an onerous job.

"If I come to your home and say, 'Okay, here's papers, we're foreclosing on your farm, you gotta move out tomorrow afternoon. I'm going to take your car. When I leave now, I'm going to take your truck,' you're going to be mad. And the only person there is me. So who do you get mad at? You get mad at me. The banker ain't standing there.

"We've had guns pulled on us. Most of the time it's just a verbal hollerin' match. They keep saying, 'Why are you doing this? Why are you being so hard on me?' I tell them, 'If it wasn't me, if I quit my job, there'd be a new sheriff tomorrow.' " The Farmers Home Administration still has plenty of delinquent loans on its books. Most cropland goes back to FmHA, the Production Credit Association, or a bank, but pastureland tends to sit idle and the government ends up owning a lot of it.

Many foreclosures, in Jager's experience, are linked to a farmer's attempt to see his son take over his land. "A lot of farmers that we find here in trouble, they got there all on their own," he says. "We find in a lot of foreclosures the son bought the father out 4 or 5 years ago at a high price and high interest. Once inflation is down, there's no way they can make it. Will the father adjust the price down? No. You see, he's got his money. The father's got it sitting in CDs or whatever; he's made a good dollar. FmHA or a bank put up the money. Now the son is going under. The father's got the funds to bail him out. But he don't. What they usually do, the son will go under and the father will come back in and buy the land back at half price. So he's still got half his money plus all his land back. We're finding that. Very few fathers are bailing out the son. They'd rather lose it and come back in. Let it go and start over.

"The son has to file bankruptcy to get out from under. It liquidates all assets and gets rid of all your bills. I mean, you're just broke. You got nothing. All land and all machinery goes. You're wiped out. But it doesn't hurt your credit." Jager says if a son goes broke and declares bankruptcy, he usually comes back and farms with the father. The farm is now in the father's name, but it's clear and there are no debts or liens against it. The son got rid of all these when he filed bankruptcy and the father bought back the farm for a percentage on the dollar. As we have seen, father-to-son continuity is such a strong tradition in rural culture, here and everywhere, that such machinations, however a violation of the spirit of the law, meet with local social approval.

Jager finds when he goes to make a foreclosure, it can turn out to be a big rich farm or a small poor farm. "You'll find a lot of the rich ones overbought in the 1970s and early 1980s," he explains. "Went completely extravagant. They put a brand-new home on the farm, brand-new trucks, brand-new this, brand-new that, Winnebagos, boats, motors, trailers, campers, motorcycles lying all over the yard. These people definitely have a problem. You know if you bought three snowmobiles and three motorcycles and a boat for top dollar 8 or 9 years ago and now everybody's foreclosing on them, they are not going to bring even half of what they cost. There's no way to come out of it. They've got to go under."

Even frugal farmers get into trouble. What usually happens on a small farm, Jager says, is somebody like Alfred Zuckerman tries to isolate himself from the economy. "You're looking at an older fellow that runs real old machinery for years and years. Which was a cheap way to go and a way to make money years ago. But now with the price of parts—they go up pret' near 25, 30 percent every year—pretty soon he's in trouble. He can't operate with that machinery. It's completely wore out. His old tractor is maybe worth $2,000, and he's got to go buy a new one. Now he's looking at a $100,000 investment for one piece of machinery. And his tractor's wore out, his combine's wore out, his cultivator's wore out. And it's time to replace them. He's looking at $250,000 just to get the machinery to do business. And he can't make it.

"More than likely his land is all paid for. He's got everything paid for. Not a debt in the world. But the repairs keep him broke. Now, all of a sudden, he's got to gear up to keep going. We're finding these guys close to retirement. They're just hanging in there until they can retire and rent out their land. Around Prairie even when it's $35 or $40 an acre, it goes fast." I thought of young Lowell Zuckerman; his father might retire, but wouldn't he face the machinery problem too?

What about a big renter like Bud Petrescu, I asked Jager. He said he finds—in contrast to the profligate spenders or frugal small farmers who fail to keep up—big renters, who own no land at all, sometimes manage to "operate big, spend big, stay up there." He mentioned one big family who only rented their land and said, "You go in their yard and you'll find three or four big four-wheel-drive tractors. And they can afford to keep up crop and hail insurance and what they need to keep them alive."

There are limits to bigness, he feels. "You can only farm so much. North a little bit over from Hamberg, one farmer, he took everything. I don't care what, whether it was good land or bad land. He just sucked it up. He just sucked up land. Now this year I see he's giving it back and giving it back. Got too big for his own good." It confirms what others said.

I couldn't imagine what it was like to have the sheriff come bill collecting. How did that go? "Let's say you bought a color TV and you didn't pay for it. Okay, the store owner gets an attorney to serve a summons complaint. After you don't do nothing, it goes upstairs for judgment. The attorney requests the court to send judgment to the sheriff. Under the law that forces us—we have no option—to go look for that TV set or something else of yours and sell it and pay the creditor. It can be your car, your savings, your checking account. We do what's called a levy.

"I come to your house and I find that you've got a car you've paid for, something I can sell. I hand you a levy. I take your car. Then I advertise it and I sell it from the front steps of the courthouse. As a matter of fact, tomorrow morning I got a sale on the front steps of the courthouse of a motorcycle. A young guy wouldn't pay a doctor bill, wouldn't pay me, and wouldn't make no arrangements. So I levied and took his motorcycle, and now I'll sell it on the front steps. If it brings enough to pay the bill, fine. If it doesn't, whatever else, we'll keep it on file."

He said the state says to always take personal property first, before real. "I'll sell your tractor before I'll sell your land. Now, there's certain household goods you can't take. Under the law if you've paid cash for it, a man's home is his castle. Nobody can take it. I can come in and sell your big stereo. Your fancy TV and VCR. But I can't come and sell the furnace out of your house. 'Cause in North Dakota you have to have heat. We can't sell the bed. We can't sell the family heirlooms or the family pictures off the wall. That's what they call 'absolute exemptions.' And it's an old law. I cannot sell so many mules, so many donkeys, so many ducks and chickens. And that law is still on the books."

Jager is the third generation of his family to live in Prairie. He has no illusions about its future. "It's going down. We got a lot of empty places on Main Street. That are just abandoned. We got one grocery store. Years ago we had three grocery stores in town, and all three had delivery service. At four o'clock they'd have the groceries right in the house. I myself, when I was in high school, was a delivery boy. When I'd come to your house, I'd put the milk in the fridge for you if you weren't home, you know. All the houses were open and unlocked."

A big employer in Prairie is the co-op–owned elevator, which bought most of the smaller elevators around. "While before," Jager says, "there used to be two or three men in these little elevators, now they got one sitting there. He buys, sells, the trucks here go out and pick it up, the unit trains take it, and if for some reason it gets busy, they'll pop a man out there to help." It was the same way with law enforcement. Each little town used to have its own police officer. Now just Prairie and Harvey do. Otherwise all you have for the whole county is Sheriff Jager, his deputy, and a highway patrolman. Since the same thing is happening all over America, many small communities are left without much protection of any kind.

Like everybody else in town, Jager says being a county seat is what keeps Prairie going. "If it wasn't for the county courthouse and state and federal offices, Prairie wouldn't be here. Look at the ASCS office. Every time they get a new farm program, they've got a new secretary or two. You take the courthouse, FmHA, ASCS, and those things, that's a big chunk of the community. And if you want to check your records, you have to come to Prairie to do it. So that automatically brings business. A farmer comes to town to do his land thing, you know, he might stop and get a part for his tractor or a new pair of coveralls. If nothing else he'll stop for a beer.

"I'd say the town is going to last as long as Hawk County lasts. And as far as the farming economy, if the years of drought get behind us, I'm looking for a bottoming out and a turnaround. For 10 years I never foreclosed a farm. Now in the 1980s I've had more than my hands full. If you go into debt, I mean, you're there. The bill collectors—which is basically me—come after you. There's not much you can say."

THE OUTSIDE WORLD gets closer. Not just in the impact of television or the way urban problems like drugs have come into Prairie. Or even in modern-minded farmers like Paul Fischer needing to keep up with events in Mexico to market their pinto beans. Prairie's main exposure to another culture until now has been to Native Americans, Sioux and other Indians from North Dakota's reservations. Now the Mexicans are coming.

What makes Mexico's villages different from all the others in the Third World is that they lie just across our common border. Short of building a 1,950-mile fence and constructing a symbolic iron curtain on the Latin American border, it is impossible to isolate ourselves from the problems of the Mexican village—too many people and too little land and water and jobs. When it comes to bridging the gap between rich and poor, the Mexicans do it with their feet. Even getting as far as Prairie.

The first Mexican family arrived in the summer of 1989 and was invited. Young Mike Hanson, helping his brother Knute grow pinto beans, got the idea immigrant labor would be better than herbicides; he phoned a college friend in Fargo who went to the Migrant Council. Some days later Esperanza Tamayo, a plump, cheerful woman in her 40s with dyed bright yellow hair, arrived in an old beat-up Buick with six children and her husband—her second, she said. Since Mr. Tamayo spoke not a word of English, it was pretty certain he was an "illegal alien," "undocumented migrant," or "wetback"—so called because they used to swim across the Rio Grande—one of the countless Mexicans who come north each year; more than 1 million get caught and sent back each year by the Immigration and Naturalization Service, what Mr. Tamayo calls *el migra*. In the late 1970s I lived for a time in Huecorio, a Tarascan Indian village in Lake Patzcuaro in the Mexican Central Highlands, where almost everybody migrated legally and illegally to *el norte* at one time or another. (Two old ladies even went up and back, quite illegally, to see *Disneylandia*.) I concluded that instead of in antiwetback outcries, the solution lay in improving lives and livelihoods in villages like Huecorio so its people could stay at home.

In the meantime, Mexico's illegal immigration is a remarkably efficient form of foreign aid (money sent back goes right to the grass roots) and helps, by providing us with cheap labor, to fight inflation. Most profoundly, since most Mexican migrants are villagers, with strong agricultural, religious, and family ties, it reinforces the rural basis of American culture.

Mike Hanson's teenage friends made jokes about the Hansons' "Mexican John Deeres." But Mrs. Hanson, a leader in the Catholic church, was very hospitable and found the Tamayos an extra-large room in the Prairie Motel, next to its laundromat, and Esperanza's family was put up there.

"I've been coming to North Dakota 18, 19 years," Esperanza told me one day as she and her family hoed weeds in the Hansons' pinto beans. The calluses on her hands, even though she wore leather gloves, showed how hard hoeing was. "We used to go to Minnesota for the sugar beets. Now we stay with one farm near Hillsboro in the Red River Valley. It's easier to get paid. It's easier to get allocation of food stamps. But we finished up early and had some time, so we came here." She said the drought made work in the Red River Valley scarcer in 1988 and 1989. They worked there in the summer, and in the fall went to the west Texas sugar fields.

"Now sugar beets is by acre. So it's up to us. Standing, sitting down, working or not, whatever. It's our job. If we hurry, we get more money. Well, we don't work too much. Just 9 or 10 hours a day. If it gets too hot, we quit earlier." In Prairie the Hansons were paying $4 an hour, which works out to be between $4 and $14 per acre, depending on the weeds. Weeding is strenuous; each of the Tamayo family walks miles and miles up and down the fields in a day.

Esperanza does all the cooking too. "We bring our lunch to the field. I got a gas burner in the room and make tortillas. They don't sell them over here. It's better homemade anyway. Flour and baking powder and water. I make tortillas every time we want to eat. In North Dakota there's not a lot of Mexican stuff to buy. Not like in Texas.

"You know what people in Texas say? They say don't come over here because they don't pay unemployment. And in Minnesota they do. North Dakota is maybe the only one. My husband, he says he'd like to stay up here all year, but he don't like the snow. I don't either. I don't know how to drive in snow. And the kids want to go to school back in Texas. We got a house in the Rio Grande Valley.

"Some of the farmers, they put chemicals. But sometimes chemicals don't work. And what I heard that they were saying, that the beans hoed by people, you know, they're better than the ones they put chemicals for the weeds. That's what they said. I don't know. I know a lot of farmers, too, they don't hire Mexican people to hoe the beans."

I heard women were better at it than men and asked Esperanza about it. She roared with laughter. "They did? Well, what I think, I think the woman works harder than the man. Because over here I'm the one that always says, 'Let's go to work!' "

"But the man is boss."

"No, that's in Mexico! Not over here." More laughter. Her husband spoke to her in a rush of rapid Spanish. He was wearing a T-shirt saying "SPECIAL FORCES." "He says that's why he wants to take me to Mexico," she explained. "I don't want to go. Well, I'd go but I'm *not* living in Mexico. I'm from Texas. I've seen Mexico and I'm not living there. In Mexico, the women don't work too much outside, as over here. The woman, she is in the kitchen. Maybe the old ones work in the fields. Not the new ones. I know because all my folks are Mexicans. They got passports. To go over for fiestas, market days is okay. But not the life of Mexico. In the evening the women go to church and the men go to the cantina to drink beer."

She said it was up to them whether they worked on Sunday. "Over here we don't go to church because there's no Mexican priest. In Crookston, Minnesota, they do a mass in Spanish and we'd drive there if we want to go to church. Movies? We rent one and see it on the VCR. Here to get groceries we have to go to Minot. Prairie is too expensive. You'd spend all your food stamps in one day if you go here."

Would she rather have work in Texas? "Stay home all the time? Not me. I like this migrant work, you know. Why? Well, because you don't come to work at the same time, go home at the same time. It's like a vacation. It's a change. You see other faces." She just closes her house in Texas. "We shut the doors and lock the gate. We have done it many years and nobody breaks in yet. The neighbors take a look at it."

Esperanza finds the biggest change over the years in North Dakota is that the fields are getting weedier. There are so many drought weeds like cochia and Russian thistle, and a new invader, black nightshade. Since the 1970s, she says, wages have hardly gone up. "It used to be every year there was a higher rate. One year it was $22 an acre for sugar beets, and the next year it was $23. We were going up. For about 10 years now we're stuck at the same price. One other thing, they ought to provide bathrooms in the field. And I think housing is a problem. A lot of Mexican migrant workers, they come and they are looking for housing. Whites don't want to rent a house. Not to Mexicans with a lot of kids."

But Esperanza likes Prairie. "We like the way everybody treats us here," she says. "If people need more field workers, I got plenty of relatives," she adds, with a hoot of laughter. "But it would be better to come earlier in the growing season when the weeds are smaller."

Poor Mexico, as the saying goes, so far from God and so close to the United States. If Mexicans are going *al norte*, even as far as North Dakota, Americans are going south. Our fates would be bound together even without the 500,000 to 1 million barrels of Mexican oil we were yearly importing in the 1980s. There are around eight million Mexican-Americans, or Chicanos, like Esperanza now, and some thirty-five thousand retired Americans live in Mexico, and in good years, about three million Americans visit it as tourists.

So that traffic between the Rio Grande Valley and the North Dakota prairie is a two-way street. If migrant workers come up, retired farmers like John Farrow's father, Lionel, and his mother, Helen, go down. For some years now they have owned a mobile home in McAllen, Texas, close to the Mexican border.

Why Texas? "Oh, I don't know," says Lionel Farrow, a retired elevator operator in his 70s. "We was pulling a trailer for 3, 4 years and we had friends there and we stopped and got acquainted. And stayed, I guess. We're only 10 miles from the Mexican border. I got a friend from Minnesota, and he's got a factory just across the line. And he says, 'We have very good help.' Of course the reason they're over there is they're only paying 50, 60 cents an hour."

The Mexican-American border is lined with such businesses, combining U.S. capital and technology and Mexican labor. Neither country talks about it much, but it has significant economic impact in both America and Mexico.

As Lionel Farrow describes it, "When they get these pieces assembled, they bring them across the line, where they got a big plant. They make the parts across the Rio Grande and bring 'em over here and assemble them. The Mexicans do this little, fine work. And Mexicans from that side, they get over here somehow and make $4, $5 an hour, some more than that. And then they slip back to Mexico."

Farrow says American border police keep picking up illegal immigrants and taking them back. "One night you might see two, three truckloads. But you can't keep 'em home. 'Cause conditions are so tough over there. Goodness. They'll get in some corner and put a cardboard box around. And that's where they live. Terrible. And they get food stamps. They'll come over here, across the river, and they get groceries and commodities and take package after package back."

His trailer park is no longer so safe. "They'll break into the park. Daylight or night. Steal a bicycle or two. Neighbor next door, they broke in there the night before we left. Took two TVs and a microwave. The lady next door, in between us, saw it and she hollered. But you don't dare do anything to Mexicans. Get the law after them and they'll be back and raise hell with you. Years ago when we first went down there, us snowbirds, and we'd go into a store, we had to wait to get waited on. But today the Mexicans don't have money, and they're taking care of us first. The Mexican waits. I mean ninety-nine-and-nine-tenths of the people down there are Mexican. I mean there's nice Mexican people."

Farrow says his trailer camp is 1,796 miles from his house in Emrick; it takes two nights on the road. "I got a diesel car and fill up here and two fillings on the way and the tank is pret'near empty when we get there. When I first had the diesel car, it was 15 cents a gallon. Now it's 48 cents in Mexico, 96 cents in Texas. I drive across and fill it up. They don't say anything about that, but you can only bring a quart of booze home a month, $1.75 for a quart of vodka. Years ago we used to drive from one town to another on the Mexican side. Now a fellow's a little leery. People get robbed. They're so hard up now, they'll do anything for a dollar."

MEXICANS COMING TO Prairie and Prairie's people going to Mexico, like farmers watching wheat harvests in Russia or China when they go to calculate when to market their own crop, or even just the way practically everybody now watches the evening TV news and talks about it, suggests, as I've said, how much broader rural America's outlook has become.

Much of the time Prairie's people go their separate ways, grouping together on Sundays as Lutherans, Catholics, Baptists, and Congregationalists. A parade or a sports event, like the Little League games or basketball and football at school, will bring a lot of people out. But what really brings all of Prairie's people together, even if only for 4 days once a year, is the Hawk County Fair. It is also one of the best ways to see how Prairie has culturally changed—and not changed—over the years.

The fair is, of course, not what it used to be. The record crowd was in 1929; just over 7,000 passed through the gate opening day. In 1989— North Dakota's centennial year, when there was an all-year Prairie school reunion at the time of the fair—the crowd on the average day was about 2,000, so the biggest change, if one compares the fair of 1929 and, say, 1989, is the drop in people. Clothing, too, is markedly different. In the late 1920s people dressed up, women wearing white and yellow summer dresses, often with big picture hats as if at a garden party, or the close-fitting cloche hats of the twenties. Men, as we see from old photographs, came to the fair in suits, white shirts, ties, and suspenders, with derbies or straw boaters on their heads. There were also occasional farmers in bib overalls and old straw hats.

The midway, 60 years ago, was much bigger, with great billowing painted banners that hailed "The Snake Charmer," "The Alligator Wrestler," "The Fattest Woman in the World." Barkers outside the sideshows would keep up a noisy patter: "Twelve United States Cavalry swords are going to be thrust through a box while this little lady is inside." You'd hear the clanking machinery of many rides and the calliope of the merry-go-round blaring away and see two giant Ferris wheels with shrieking riders waving to their friends. The noise was terrific: concessionaires shouting, "Come and eat! Ain't you hungry? Roasted peanuts! Hot dogs! Hamburgers! Popcorn! Cotton candy!" Men would be selling squawking toys and rubber balloons on sticks and pink plastic Kewpie dolls and little windup clowns with canes. A voice would boom over a megaphone from the grandstand, "Ladies and gentlemen, tonight you will witness one of the great death-defying feats ever attempted in these United States!" From the *Hawk County Free Press,* Prairie, July 10, 1929:

> There are plenty of thrills and comedy in the free acts this year. The Flying Fishers, aerialists, headline the program. Daring feats high in the air are performed by this troupe of three men and two ladies. Charles Fisher, their leader, does a real thriller when he makes a somersault and change in mid-air with his head in a gunny sack. The

Two Crazy Sailors have a comedy tumbling act but the big laughs go to the Gold Dust Twins. . . . The John Francis Shows on the Midway offer the best entertainment of any carnival ever here. They were so crowded for room it was impossible to set up all their shows and rides. . . .

The machinery exhibit attracts a great deal of attention. New types of labor-saving devices for harvest especially and shiny new threshing machines are lined up for inspection. . . . Of course, it wouldn't be a fair without Rube Liebman. He is here this year with a new line of chatter but it is the same old Rube, on the job all the time. A loud-speaker system gives him a better chance to make all the announcements from the platform. . . .

It was just 3 months before the Crash plunged the country into the Depression. The county fair was the biggest Prairie had held or ever would hold. There were over 400 livestock displayed—157 cattle, 118 hogs, 23 Clydesdale and Percheron draft horses, and 97 sheep. Every pen in the poultry show was taken. The 4-H building was brand-new. At the horse races, Rube Liebman—a standing comic and emcee from Minneapolis who came back year after year—wore a goatee, a top hat, and a star-spangled suit like Uncle Sam. He joked about the new technology, telling the crowd over the just-installed public-address system, "Modern times are here to stay, folks. Yes, sir-ee. How many of you heard of a radio 5 years ago? Or saw an aeroplane 10 years ago? Or owned an automobile 15 years ago?"

Today, looking back, it seems amazing radios, planes, and cars all came so fast. At the 1989 fair, Rube Liebman's successor, an affable local auctioneer, Otto Schultz, who wore his habitual Stetson and cowboy shirt, was still going on about new technology. He told about a trip out to Los Angeles: "The pilot comes over the P.A. smartlike and says we're flying at 33,000 feet. And don't you suppose we hit one of those air pockets and all of a sudden we're down to 20,000. That good-lookin' lady in that uniform, she comes around with some lunch and I said, 'Look, you might just as well dump it in that bag. That's where it's going anyway.' " The humor hadn't changed in 60 years.

When he got to Los Angeles, Schultz said, he found his suitcase hadn't come. "So we left, drove 30 miles through that traffic to get to my sister's house, and we barely got there and the phone rang from the airport. And they said, 'We found your suitcase. It's in Hawaii.' 'Hawaii!' I said. 'It's happened again! I've never been in Hawaii in my life, but I've got a suitcase that's been there twice.'" Once on a flight, Schultz said, the stewardess told him, "We got some good news and some bad news." He said, "What's the bad news?" "We're lost." "What's the good news?" "Well, we're making good time." Corny as ever.

Prairie's old timers like to remember how the fair used to be. Bud Petrescu's mother, Anne, says when she was young, the Hawk County Fair was the biggest even for 100 miles around. "People saved up all year just to go to the fair. The grandstand was so packed, you couldn't even get in if you weren't there early. And, oh, God, those big velvet hats in July. White was very, very popular. They used to keep the grandstand real clean, sweep it off and scrub it. And everybody wore white. The men wore those stiff white collars which detached from the shirt."

Owen Rhys says three big half circles of Model Ts, open roadsters, canvas-roofed and windowless, but a few Model As too, would park facing the track, grandstand, and judge's stand. "They'd park Model Ts all around the track and sit there and watch the races. The women in white dresses, the men in suits, heavy-looking suits. My God, it must have been terrible. So hot. A lot of farmers years ago would put on a pair of clean overalls and a Sunday suit coat and come to town. I don't know why. Maybe because you had to deal with horses, hitch' em up, and overalls, I suppose, were easy to wash."

Phil Spengler, a long-retired grocer, recalled the carnivals used to come by train, in twenty or thirty boxcars. "They'd ship in at least a dozen sideshows and the rides by train. All kinds of stuff. Oh, golly, I can't tell you. They had reptiles, hula dancers, and they always had girlie shows up by the horse barns. Yeah, sure. Women went with their husbands and then the barker would say, 'All the women have to leave now. Men can stay if they pay 25 cents more.' And they'd do a striptease. Oh, those carnival people were a wild bunch. I tell you, some of those athletic shows, those wrestlers and boxers, they had some pretty rough characters.

"Oh, we had very high-class entertainment in front of the grandstand. The big races were the harness races in those days. Rube Liebman used to come many years. He had a tailcoat and he had maybe fifty or a hundred different kinds of buttons on it. And he had whiskers. And an old suitcase that something was dragging out of. For a time they didn't have a loud-speaker, and he had one of those horns, megaphones, and he had a voice you could hear for a mile. It was like a bull.

"Oh, and some of those aerialists. This one diver, not long afterwards, he got killed. He missed and hit the side of the tank. Some of those people performed without a net. I didn't even like to see it. . . . The Gold Dust Twins were two black boys from Chicago, I would say about 17 or 18 years. Both played banjo and sang and tap-danced and they told jokes and it was really a scream."

Gebhart Bauer remembers how they used to have horse pulling contests with great big draft horses. "And they'd have relay races, chariot races, mule races, Shetland pony races, ladies' races, Model Ts, free-for-alls. There was always the old tug-of-war. That was one thing the farmers had to win even if they had to cheat a little. If it went on a little while, the town guys' arms would ache and their legs would ache. The farmers were used to work and were stronger."

He recalled the alligator wrestler was a big joke. "All he'd do is rub its belly and that alligator was the happiest thing in the world. In the wrestling tent, three or four guys would travel with the show and wrestle local guys for maybe $20. It was a very tough battle. They wrestled for blood. Those old guys in the carnival weren't about to lose their 20 bucks."

Karl Fischer remembers the wrestlers vividly. "Two men stood on the platform outside. They'd take on anybody. And us kids would just wait for somebody local to get up and beat 'em. And sometimes some guy who'd never wrestled in all his life, he'd say he'd take one on for $20 or whatever it was. And then the show would open and we'd spend our dimes to go in and yell for the local guy. The girlie shows, you'd go in there with your wife and they'd say, 'Ladies, if you want to go out the back way here, it won't be long and the men will meet you out front.' It was just a strip. They took off everything. And then some women in town would hear about it, and they'd put some pressure on the commissioners. And maybe the last night of the fair, they'd have half an hour left of the show and the sheriff would come in and close it. Just to say they'd closed it, you know."

Owen Rhys says that in the old days there was always what was called a bowery dance. That is, a dance on a wooden platform covered with a canvas roof. "It wouldn't take a heavy rain, but a light shower, it would cover 'em pretty good. People came from all over. The dance floor used to be jammed. Oh, there was the odd fight. I suppose it was a little too much liquor. You know, that home brew, when that hit 'em, that hit 'em like a hammer. If they were a little grumpy to start with, it wouldn't take much." Years ago the main worry was driving home. "Those dirt roads, if we had a heavy rain, the car would slide all over. Unless you got a good rut started.

You could follow a rut. Yeah, until they got gravel on the roads, there were a lot of cars in ditches."

These days there is just one Ferris wheel and the wrestling shows and girlie shows are long past. The 4-H poultry exhibits were down to three this year and five last year; there used to be more than fifty. The cattle barns are half empty, but there are more racing horses than there were in the old days. In the women's hall, wooden and white painted and going back to the 1920s like all the Hawk County Fair buildings, there are still big heaps of vegetables and quilts, embroidery, hooked rugs, and cut work, and all kinds of flowers.

At the grandstand there are 220-yard races for ponies and quarter horses, five-eighths-of-a-mile opens and consolations, and three-quarter and mile-long derbies. Hawk County prides itself on its thoroughbred racing. Just about all the jockeys are young Sioux or Ojibway from the Turtle Lake and White Earth reservations.

The grandstand steadily fills up. The gathering sense of excitement at the races hasn't changed. Otto Schultz's voice crackles over the loudspeaker, "The boys are ready. 'Boots and Saddles' will be played. Now folks. . . ." Schultz keeps up a running commentary over the brassy sound of the band. Looking around, one sees so many familiar faces. Way up high, where the grandstand's flags and red, white, and blue bunting flap in the wind, are the Schumachers; she is fanning herself against the muggy, sweltering heat. Behind them the sky is a faint greenish purple and there are faint rumbles of thunder. Not far below, sitting on a blanket, are Owen and Alice Rhys. Elsewhere in the crowd, one can spot the Fischers and the Whites and so many other families. Mostly the generations sit apart, though the Zuckermans are all together. One recognizes the Petrescus, the Petersons and the Hansons, the Farrows and the Muhlenthalers.

"Now folks, this is the open one-mile derby," says Otto Schultz, and everyone is intent on the race. "Just like the Kentucky Derby on a smaller scale here in Prairie. You've got seven horses. Number one is Shawnee Red. Owner is John Ness of Wessington, North Dakota. And John Slater is jockey. Number two is Quick Victory. . . ." After he names all the horses, Schultz reminds everybody that it is the last night of this year's Hawk County Fair and they are going to close with a talent show. "It's amateur night," says Schultz. "Right out here at the grandstand."

Just before the race, a roar goes up from the crowd. One horse, Quick Victory, has got its nose stuck in the starting gate. Men rush over, climb the gate, and release the frantic horse, which is badly frightened but unhurt. The race begins. "And they're OFF! Infinity Factor's in the lead, with Rooster Charlie coming up on the outside. . . ." The horses no sooner reach the back stretch when Quick Victory, still skittish, throws its rider off. The jockey, Bud Goodwin, who turns out to be John Running Horse's nephew, lies in the dust. He is still and doesn't get up. Otto Schultz shouts, "Oh, oh, they need help across the track there, fellows." People in the stands jump to their feet, trying to see. Two men run to a parked ambulance and in a minute are driving it over. Other men are running across the grass. Goodwin's horse turns and runs back toward the gate that goes to the barns. The other six horses, oblivious to the accident, race on.

"Quick Victory has fallen and lost his jockey!" Schultz cries over the loudspeaker. "The rest of the field is coming. Quick, get 'em off the track, boys! Get 'em off the track! Those horses are coming round!" Just in time, the men reach Goodwin and carry him to safety on the grass beside the track as the pack of horses hurtles by. One of the men is Prairie's doctor, and word soon passes that Bud has broken a collarbone. But now the horses are thundering into the home stretch and the crowd is on its feet, shouting and screaming. Schultz's voice rises above the uproar, "And it's . . . Infinity Factor! Infinity Factor is the winner! One-forty-six-oh-eight is the time. Rooster Charlie's second. Ship Wrecked is third. . . . Folks, that's the 1989 one mile derby, $450 purse. Here's the owner to get his picture taken. I see the whole family's coming to get in the picture. That your son? How come he's so good lookin'? Give 'em a big hand, everybody!"

The storm hits just after the races end, one of those summer thunderstorms rolling in from the northwest, with rumbles of thunder and lightning flashing every minute. All at once, a squall of wind tears over the grandstand, with such violence several men grab their caps. A black cloud sweeps down, and with it comes the rain.

Everybody scrambles for the Catholic food tent as the first drops spatter down. The Catholics are serving up their yearly roast chicken dinner with mashed potatoes and gravy, string beans, fresh rolls, and all the trimmings, and plenty of homemade pies, a la mode if you have any room left. And steaming urns of coffee, the weak Scandinavian kind North Dakotans like so you can drink several cups. Nobody wants to miss the dinner, and it won't be long until the talent show. Father McNeeley, the Catholic priest, and Mary Pritchett, Knute Hanson, and a dozen other Catholics are kept busy helping to dish up trays. The old tent was rebuilt in 1987 into a large roofed wooden building, though everybody still calls it the Catholic tent out of habit. It is quite a triumph over the Lutherans, bigger in number, who gave up their food tent some years ago.

The new building has become a gathering place for Prairie's middle-aged at the fair. The elderly tend to congregate more at the museum, which has its own coffee maker. The young go, as they always have, right to the midway, such as it is nowadays. Others crowd into the beer garden under the grandstand, which gets pretty noisy. Sheriff Jager and his deputy come by every so often to keep an eye on things.

Summer storms, if they bring hail nearby, can mean a sudden drop in temperature. By the time everybody takes their seats in the grandstand again, it feels more like October than July, and families huddle together, wrapped in jackets or blankets for warmth. A wooden stage has been set up for the talent show. There is a piano and a microphone.

The wife of the fairgrounds custodian comes on first, a tall, nervous-looking woman in a cotton dress. She says she will sing a revival hymn. It is "Keep Your Hand on the Throttle and Your Eye on the Road." Her voice is hesitant and not always audible, but she is warmly applauded. Everybody gets a good hand. Charlie's barbershop quartet, all dressed up in red vests and black string ties, sings "The Streets of Laredo" and "The Colorado Trail." Their harmony is good and the crowd brings them back for an encore, "Rise Up, O Man of God." The quartet has to come on early, as Charlie has to drive up to Harvey for a rehearsal of *Oliver!*—he will be Bill Sykes.

Andy and Ingrid Rynning come on third and sing "Tumbleweed" again, as they did when we met. After them a very old, bent-over, white-headed man plays "Red River Valley" on his harmonica. Beyond him, across the racetrack and the fairgrounds, now rainswept and looking very wet and green, is the open North Dakota prairie itself, looking just as flat and bare and empty as it must have looked to the Sioux. How vulnerable your little society is, it seems to say. One tries to imagine how lonely and isolated the first settlers felt under such an enormous cloud-heavy sky.

Lois Phillips Hudson in her *Reapers of the Dust* says the North Dakota sky has "the kind of height that only the sky of a prairie or a desert or a sea can have; it makes its own boundaries, its symmetry never spiked by the reaching of tall trees, never crowded by the peaks of mountains." Hudson tells of "the loneliness of great spaces" and says, "In that enduring loneliness I might have existed through centuries of freedom and bounty, when the grass rose to the shoulders of the buffalo and the grass and buffalo held each other against wind and drought."

For now the sound of piano, harmonica, guitar, and human voices, loudly amplified, muffles the shriek of prairie wind, even if you can feel its raw chill go right into your bones. In spite of the cold, nobody goes home until the talent show is over.

CHANGE NEVER COMES easily. For instance, the Catholics practically took up arms over the matter of the food stand at the county fair. When they built the new building, the younger generation decided it would be a lot more efficient to serve cafeteria style, with everybody taking a tray and moving in a queue. The old guard said they'd always taken orders and brought the food right to the tables in the good old-fashioned way. Why change? "That was a knock-down, drag-out battle," says Father McNeeley, who did his best to stay neutral. "The old guard lost and the new way proved to be more efficient. But not without some bloodshed."

Father McNeeley feels Prairie has always been a social kind of place. He compares it with Norway, his other parish in the next county to the north. "My Norway people are farmers. Going back to the early settlers. There were Irish Catholics there before there was a Norwegian in the place. Now here in Prairie there's a completely different feel. My parishioners are mainly townspeople. From elsewhere. Only a few names go way back. Prairie was not a Catholic-settled place. They moved in later. Here the church is a social institution. I've got a whole force of energetic young mothers. They're a ready work force. They've got time to do things, they've got the energy, and they've got the willingness. They take something and make it prosper."

Maybe they are *too* social, he says. "I wonder sometimes if their motives are really pure. They'll compare themselves with what the Lutheran ladies are doing. It's kind of a social club. That we have to keep up with. We should have a piano too. See what they have. And we got a piano. This town is very social. They've got the golf club out there. Ladies' night is sacred. Men's night is sacred. It's frustrating. Oh, it's not frustrating to accomplish projects and tasks. It's kind of a joy. They *do* it. But I don't want to preside over just a social club. I don't want to just 'play church.' "

Father McNeeley, with his freckled Irish face and sandy-colored hair, looks to be in his mid-30s; there is a youthful earnestness about him. "Norway, depressed as it is," he goes on, "there I can point to some real, authentic religious experiences. People that know the value of openness to God."

He says he is worried about Norway. "It's richer down here. Better soil. There's just a pall of heaviness over people there. Not everybody. We've got a group of six or seven young farm couples; they meet twice a month. It's a small mutual support group, trust, sharing. We've got the young Lutheran pastor and his wife in it too. And the Rynnings. Mostly it's young farmers from South Norway. There's so much life in them. They've all been hard put."

Father McNeeley says he keeps asking himself what makes Norway so different from Prairie. "Ever since I've been in Norway, it seems like every year somebody gets crucified. It's usually centered on the school. I mean, every year they seem to pounce on somebody. And get everybody up in arms. The whole place, it just gets divided up. Every year something seems to heat up controversy. Calumny. It's vicious. It's depressing. I feel so much heaviness up there. Someone in that community has got to stand up and say, 'Do you think we could be nice to each other this week?' I'm talking about the town itself. Those farmers around South Norway don't get involved in it."

He speculates the small Norwegian community, mainly descended from its original settlers, has too many families that are closely related. "Maybe it's too inbred. There is a kind of bright light in a community when you've got some outside perspective. Prairie has lots of ties with the outside. Charlie Pritchett spent a long time in the Twin Cities. One of the bankers was a White House military aide under Nixon. One of old Judge Johnson's grandsons got his degree at Harvard Law School a couple of years ago. It's that outside influence that really seems to stabilize a community."

Father McNeeley says he used to have a parish in Calvin, a farming community north of Norway. "They had an aerospace engineer who'd come back to farm and a civil engineer who'd married a Lebanese journalist in Cleveland and brought her back here. In Norway I don't see a lot of that."

The priest, who joined the army since I met him, eventually decided the paradox to him was that while he liked living in Prairie better, in Norway he'd had more success sharing his faith. "There's people who have really come into the essence of religious experience. It's not really centered in the parish church. It's just there. In individual lives. I consider my time in Norway well spent, not being at church but with these people."

To go to its Lutheran, Catholic, or Baptist churches Sunday morning is to see Prairie at its most hopeful. After years of dwindling congregations, and more and more old people, church attendance has grown once more. Most of the gain has come in young couples, from farms or town, with small children. Indeed, enrollment at Prairie's primary school, as I said, after steadily dropping for years, surprised everybody by going up to 140 in 1990, up from 88 in 1980. Charlie Pritchett estimates fully three-quarters of Prairie's people go to church or Sunday school every week, except in summer baseball season. The official North Dakota church attendance figure is an amazing 77 percent.

If, as modern anthropologists say, religion is the core of any culture, this rural American religious vigor cries out for explanation. Generally in the West—Britain, Germany, and the rest of Europe—the number of church-going people has dropped to 3 to 5 percent of the total population. The National Council of Churches says 60 percent of Americans are active members of churches and synagogues, a figure that has stayed fairly steady over the years. Other authorities put it at about 45 percent. There are also now 2 million to 3 million Muslims in America, about 500,000 Hindus, at least 10 million to 15 million Hispanic Catholics, and a lively black African church.

What is changing is a steady loss of members by mainline Protestant churches—Lutherans, Methodists, Presbyterians, Baptists, Congregationalists—and a rapid growth by fundamentalist churches—the Assembly of God, the Evangelical Free Church, the Baptist General Conference, to name a few. North Dakota has so many Lutherans—more than 500 of the state's 1,695 churches are affiliated to the American Lutheran Church—the loss to fundamentalism is relatively small but baffling.

Some argue that in times of rural crisis, the clear and simple answers offered by the fundamentalists offer a sense of security. Pollster George Gallup, Jr., argues the whole country is suffering "a deep spiritual malaise." He defines this with such specifics as cheating on taxes, extramarital affairs, fraudulent telephone charges, pilferage costing department stores $4 billion a year, and defaults on education loans by many students. Gallup, an Episcopalian, argues people need to learn how to bring biblical principles into their lives.

My own feeling, based on experience, from living in Third World villages to writing about modern Britain, is that religion gets weaker the more scientific and urbanized a society, any society, becomes. In British society, 90 percent urban, it is not surprising just 2 to 3 percent go to church. In Bangladesh, one of the world's least urbanized, least scientifically educated societies, it would be hard to find anybody who was not a true believer, either Islamic or Hindu. North Dakota, being one of America's most rural states, is also one of its most religious. It stands to reason. Its society is based upon fairly fixed family and community ties, its people live close to nature, and their lives are in tune with the seasons. The agricultural moral code and ethic of mutual self-help are still rooted in day-to-day economic reality.

In 1987 when I was in Prairie, Billy Graham came to Fargo for a 3-day "crusade." He attracted over 20,000 a night, a number equal to about a third of Fargo's people. But many, probably a majority, did not come from the city but from little farming communities all over North Dakota and eastern Minnesota.

"Some people believe that you can sow your oats all week and go to church on Sunday and pray for a crop failure," Graham told his farm-minded listeners. "Be sure your sin will find you out." Graham chose a biblical text suited to his largely Lutheran audience—quotes from Paul's letters to the Romans and the Galatians saying that what matters is faith and faith alone. So many of Prairie's people made the 175-mile trip to Fargo to hear Graham, the Baptists cancelled Sunday church services that week. Not everybody was enthusiastic. Jack White told me, "I wouldn't walk across the road to hear Billy Graham."

Rev. Andrew Anderuud, Prairie's Lutheran pastor, was one of those who stayed home. "Billy Graham is probably the finest evangelist of our era," he said when I asked him about it. "If Lutherans oppose him, it's because they oppose decision theology. For us belief is by the grace of God." Reverend Anderuud blames the falloff in mainline church attendance to their liberalism. "So many sermon themes are oriented toward politics or psychology, instead of the scriptures. You always have the social gospelers pooh-poohing evangelism and the evangelists pooh-poohing the social gospel. But I think there's no comparison between Billy Graham and these TV evangelists. All that financial mismanagement and obvious greed."

Anderuud interestingly blamed many present doubts on the 1960s. "That whole emphasis in the cities on questioning authority and throwing out old values and doing your own thing. That sixties generation." It was a time, of course, when many young people—I was just in my 30s myself—tried to live according to freedom of individual choice, which meant throwing away a lot of the old restraints, patterns of obedience to authority and religious convention.

"Then youth is a time of exploration," Rev. Anderuud said. "There's a time when they try to find themselves. And sometimes they park God to one side while they explore other areas. We see a lot of young couples come back to church once they have children. Couples we haven't seen in 5 years. Back now that the kids are starting Sunday school. We have a whole flood of little kids right now."

People do find, I think, once individual freedom is explored, that they are happier if they do live in groups. By that I mean a small community that offers social support and instills self-confidence bred of its own sure sense of identity. This has been going on a long time. I thought about this kind of confidence one day when I happened to meet Emma Muhlenthaler on Main Street. Mrs. Muhlenthaler, a farmer's widow I'd met on a previous visit, had just returned from her second visit in 10 years to a leper colony in Africa. Surprisingly for such a small town, Prairie has two missionaries in Africa, both Baptist and both nurses; Mrs. Muhlenthaler's daughter has spent 18 years in Cameroon, and the other missionary has been in Nigeria even longer. It was quite a trip, Mrs. Muhlenthaler said; she was still getting over her jet lag. From Douala on the coast, it had taken 6 hours in an old Land Rover on a paved road, then several hours more over dirt tracks in the bush. Her driver, coal black with tribal scars on his face, told her if it rained, they might get stuck for hours, even days. She said in places the ruts were 2 feet deep.

Yet, hearing her tell her story, it was evident Mrs. Muhlenthaler hardly gave thought to personal discomfort or danger, so absorbed was she in Africa. The starved greenery, red soil, and hot sweet smells were just as she remembered. They were still burning grass to clear the bush for planting. And always the sound of drums, somewhere way off, faint but never still for long. One change she noticed was that most of the mud huts no longer had the old thatched roofs of elephant grass or palm fronds; they had aluminum sheets instead. People still slept on mats and ate cornmeal mush with hot sauce ("I took one little taste and my mouth just burned") and did without refrigerators, television sets, or vacuum cleaners, though there was talk of electricity coming in.

Somehow to Mrs. Muhlenthaler, product of farm, family, Baptist church, pioneer spirit, and small-town togetherness, Africa did not seem so strange, though she remarked she saw just one tractor and one woman in slacks the whole time. She says her daughter feels she is safer in the African bush than she would be in some big American city. "Because everybody knows who the missionaries are, and they're going to make sure nothing happens to them. I have an idea she's going to make that her home."

Mrs. Muhlenthaler found that when they went down to the leprosy village, everybody sang songs for her. "At church when the choir marches in, they kind of have a tempo or beat. And when they take the offering, they kind of dance up. If people here saw it, it would look funny, I suppose. But they're showing their joy. Those people who do not have anything, they want to give you something."

They gave her a dozen eggs and some kola nuts. When her daughter said this was her mother, the lepers would cry, "Your mother!" Mrs. Muhlenthaler said, "They just have to shake your hand and welcome, welcome, welcome, they're so happy. Oh, they kinda let out a cheer, you know, because I was not afraid to go around and shake hands with everybody. And some of 'em, you know, just had a stub of hand left, and they'd reach out to you anyway and you'd take 'em by the wrist or something like that."

PRAIRIE'S FARMS, IT seems likely in the 1990s, will get bigger and fewer, but not dramatically so. The community's future, as Charlie says, really depends on whether Prairie loses so much of its business it can no longer bring people into town for *anything*. And that is unlikely to happen as long as it stays a county seat. And, as I say, the future of America's urban culture depends on keeping its rural base, that is, on keeping enough Prairies and the farms around them alive. The effect is cultural; the solution is economic: keeping as many people as possible on farms.

We heard several farmers say that anybody going into farming in the 1970s had to bid against the greedy ones. If a farmer had, say, 6,000 acres, and another 160 acres were for sale, he would be out to get that quarter. By 1990 you no longer found that. A quarter might be worth $400 an acre, but the farmer selling it would have to get $600 an acre to pay off his debts.

Unable to get as high a price as they need, a few smaller farmers file for the complete liquidation Sheriff Jager talked about, commonly referred to as Chapter 7, its place in the bankruptcy law. Then the land goes back to the original owner and a court decides what to do with the rest. Larger farmers go through what they call Chapter 11, the way Andy Rynning would have gone if the Farmers Home Administration had not restructured his loans. Under it, a farmer's creditors are put on hold while he comes up with a plan on how he feels he can keep going. If his plan is approved by the court, the farmer amortizes out some of his payments and decides which creditors get how much per year. Some lawyers argue that Chapter 11 doesn't work for farmers who are too far in debt. If grain prices keep going down, there is no daylight ahead and they are just buying time. Some old-timers warn that if you farm 40 years, maybe 2 of those years you'll really make money.

Charlie doubted that as much as two-thirds of the land around Prairie was owned by absentee landlords, but said he'd believe it. Bud Petrescu bet it was up to three-fourths. The official estimate at the Agricultural Stabilization and Conservation office in Prairie turned out to be 60 to 65 percent of the county's land was rented, with 30 to 40 percent of rent paid on this land going to absentee landowners. It could be a retired couple who spend half of the year in Texas, like Lionel Farrow and his wife, or people in a distant city who just inherited their land, like the two brothers in California Bud Petrescu mentioned.

This farm-to-city flow of money is a big drain on rural America. Yet it seems built in. In a city one saves in assets: something like a house, a CD, stocks, bonds, savings. Farmers, when they save, buy land. Land is their retirement account. This is why a place like Hawk County has so many younger tenants. The drain is worsened by many farmers themselves, who just grow grain and no longer keep animals and take themselves to Arizona, Texas, or Florida every winter. Their time is free, it's blizzard season and 10 below, and they go.

We found many young people in Prairie entering farming now. This is probably no accident. The end of the 1980s, drought and all, were in some ways the best years in a long time to enter farming. Land prices were down—at this writing they still are; what was $600 an acre in Hawk County in 1981 and $350 in 1987 was about $400 in 1990—and with so many bankruptcies and auctions in 1986 and 1987, the prices of used machinery were down. Even rents were down, the county average dropping from $28 or $30 in 1987 to an average of $26 in 1990. And guaranteed government program payments were fairly high. A young farmer, like Knute Hanson or the Schumacher boys, can look at that and think, "Well, this is the payment I'm going to get. That's going to pay my cash rent, and at least I'll get that. Hopefully I'll also get a decent crop." He figures he is not going to get rich, but he and his wife and probably a child or two will be able to live.

A good-sized family farm in Hawk County, I think we can reckon, is around 1,000 to 1,500 acres. Paul Fischer and Harry Schumacher, like Nels Peterson, were held up to exemplify good farmers. They easily could have come as close to going broke as John Farrow or Andy Rynning—or Ernest Waldenberg, if he needed a $300,000 bailout, for all of that. It is true they started out owning much of their land, a great advantage, but in the late 1970s they could have done what others did and mortgaged some of their land to buy three or four quarters more, 480 or 640 more acres. The attitude of both of them was pretty much: "What I've got, I've practically paid for. I'm living comfortably. Why do I want more?" Both Fischer and Schumacher are hard workers. They don't get their crop off and buy campers and go off to Arizona, as so many of Prairie's farmers do. Their farms are their businesses. They may take a day off now and then, but they pretty much keep their nose to the grindstone. And like Nels Peterson and others who raise beef cattle, diversity enables them to keep their heads above water if crops fail. Fischer has his pinto beans and seed plant, and Schumacher his turkey sideline. North Dakota is not as dependent on wheat as it used to be.

When it comes to the father-son continuity issue, both Fischer and Schumacher are bringing their sons into farming with leasing arrangements. Both fathers had the foresight and ability to expand their farms sufficiently so that when the time came for the son—Steve Fischer or George Schumacher—to marry, settle down, and farm, there was enough land so it could be split, amoebalike, into two farming units to support two families. In both cases, the son rents part of the land and the father lets him use his machinery. Paul Fischer shares his 2,000 acres with Steve, Schumacher his 2,800 acres with George and Kurt, the second son who is still going to college and farms part-time. Both fathers had accumulated enough land ahead of time to swing it. Jack White, with just 750 tillable acres, had not.

How spread out and fragmented can one farm be? Chuck White found renting land just 8 miles from his father's farm was too far, though many farms are at least that scattered. Two-way radios and other new communications equipment make it possible to easily link together tractors, combines, cars, pickups, and homes. About 20 percent of Prairie's farms have them. Some move machinery 30 to 40 miles, though Knute Hanson told me he was fed up hauling it such long distances. What makes a distance too far seems to depend on whether a farmer has the right kind of equipment, the big trucks and trailers, to move machinery on a highway easily. That itself is a big investment.

Are rents, as so many younger farmers say, too high? Again it depends on each individual case. Take a 1,000-acre farm with an 800-acre wheat base. Under current support programs a farmer can plant 72.5 percent of that, or 580 acres, in wheat. If he gets Hawk County's average yield of 30 bushels per acre, he will get a payment of roughly $36,000. But say an identical-sized farm across the road has only a 400-acre wheat base, because it was heavily planted in beans or sunflowers in 1981, the year the wheat bases were determined. This farmer can plant only 290 acres of wheat, so his guaranteed government payment is just $18,000. Obviously anyone who rented the second farm would want to pay less.

Every owner really has to sit down with his tenant and negotiate. The going rate for an acre's rent for several farmers interviewed was $30. I feel this is more than most renters should be paying. Some no longer make money. One man told me, "I got the most wonderful letter out of the blue at the height of the drought. My landowner lives out west and he wrote me, 'We're going to lower your rent $5 an acre. I've been watching the drought on TV and I realize times are tough.'" I asked Steve Metzger, who runs Prairie's Agricultural and Conservation Service office, about landlords. "We get all kinds," he said. "Some will come in and complain, 'My tenant hasn't paid me for last year's rent.' That happens time and time again. Then you get the landlord who's greedy. Say he has a neighbor who is a good farmer and willing to pay him a fair cash rent. But some fly-by-night guy from 50 miles away comes over and offers him $10 an acre more rent. Maybe the owner doesn't know him from Adam, but the guy says, 'I'll give you half of it now and half later.' Well, they get the half but never see the rest.

"Then you might get—we have to chuckle about it—Joe Smith here who owns this land and rents this one. Two separate farms. So he brings in his production records to prove the yields. And what do you know! It just seems like most times the land he owns, the farmer proves a 35-bushel yield. But the land he rents on crop-share, he only gets 30 bushels on it. It's like he took a little more grain off here and put it over there.

"We even get landlords who come in and say, 'Doggone it. My yield should have been as high as his. My land's better than his. How come he can prove a higher yield?' Well, who knows? A yield can depend on so many things. Even what day you planted your crop. In a lot of cases, it's just natural to do a little better job farming your own land than rented land. You know, you own it. You're going to spend a little more time and money taking care of the weeds. Picking up rocks. It's just human nature. You're going to be closer to it. You probably get that in first and then go do the rented land next."

What nobody in Prairie seems to question is the idea of making it mandatory to deliver the highest possible yields per acre to be able to show a profit in a good year and be eligible for support payments in a bad year. It means everybody has to scramble to continuously boost their yields at a time of surplus production and environmental worries. You would think income, not production, supports would make more sense.

Looking ahead, if the drought years are really over, Prairie will still face a bleak future if (1) government payments continue to keep going down, as it looks like they are going to; (2) foreign markets don't increase, an iffy situation; and (3) oil and hence fertilizer costs go up.

In 1988 fertilizer prices actually came down. But the August 2, 1990, Iraqi invasion of Kuwait has wreaked havoc on the world's economy. Higher fuel, fertilizer, and feed prices in the immediate future also raise troubling questions about American farming's longterm fate. Is a post-oil era, long thought to be at least a comfortable century's distance away, looming up much sooner? Foreign markets are also impossible to predict. Russia, even if it doesn't go completely to pieces, needs imported grain to improve diets. The Soviet Union had a bumper grain crop in 1990, even as high as 250 million tons, but some, maybe a good deal of it, went unharvested. It stands to reason that Russian farmers, once they have secure lifetime leases on their land, will have the incentive to produce more, and Soviet food production eventually ought to shoot up, just as China's did when it gave land back to the peasants. In the meantime, anything could happen. If the slide toward chaos continues in Russia, American farmers might even find themselves providing famine relief. In the rest of the Third World, unchecked population growth keeps eating up gains in food output. But reports from India and Africa are mixed. The 1989 world grain harvest of 1.66 billion tons was exactly the same as it was in 1984, and in 5 years we'll have 440 million more mouths to feed. Yet with Russia, China, India, Brazil, Indonesia, and a few other giants at last doing the right things to grow more of their own food, the future export picture is hard to predict.

I'll quote expert opinion on this and other policy issues in a concluding section. Here just let me say that after 50 years of Congress saying, "We can't keep pumping money into farm bills," with the deficit it is starting to look like it might mean business. As usual, the Republicans will talk about wanting to see more of a free-market system, the Democrats stronger price supports. But nobody wants to go back to paying the $26 to $27 billion a year in price supports of the late 1980s.

Robert Harris, down at Prairie's First National Bank, feels confident the farming situation will improve in the 1990s. He says for 15 years land around Prairie was inflating at 10 to 15 percent a year until about 1981. The banks didn't really get into trouble until 1985. "When land sky-rocketed, our local farmers upgraded their standard of living, they upgraded their equipment, they upgraded everything. They refinanced lots of debt against their land. And then lots of lenders started saying, 'Uh uh. We're not going to keep loaning against pie in the sky.' So people aren't able to borrow against the value of their assets anymore. Standards of living fall. The guys who refinanced lots of debt against their land face a terrible, terrible problem. If enough of them can retire debt, the 1990s will look a lot brighter."

So many feel the worst of the bankruptcies and auctions is over. It looks like most of the farmers in bad trouble have already sold out, maybe 80 percent of them. Harris agrees. He says that in the late 1980s many farmers were able to retire some debt and gain a better grasp of their financial standing. "But I'd still say, not to be harsh but to be realistic, that 10 to 20 percent of the farmers left in Prairie are not going to make it. They're able to get through the drought and hang on for a few years by some of the safety nets. Society will need to decide: do you allow them to hang on or do you allow them to fail and give young farmers the chance to buy their property and farm?"

It is a hard question. One of my main findings in Prairie, I told Harris, is the role generosity plays in father-son continuity. "It's extremely important," he agreed. "If parents are generous, their sons and daughters have respect for them. One of the sad things is where the son thinks he knows everything and the dad thinks he knows everything and they can't talk and split up. It happens a lot."

We leave Prairie, turning now to Crow Creek—our other very different, half-urban, half-rural, half-commuter, half-farm community—hopeful that the turn of events and some form of government help to farmers will keep enough of the Prairies of America going to provide our culture with its indispensable rural base. Robert Harris gives part of the reason: "I think if the government dropped price supports tomorrow, even the best farmers here—your Fischers and Schumachers and Petersons—couldn't make it. It would just be an upheaval. Everybody—big farmers, small farmers, small towns, rural businesses—would be badly hurt. The social impact of that kind of upheaval would just be terrible."

But not as terrible, I would argue, as the reflection of that upheaval in our decultured cities.

# III

## Crow Creek

CROW CREEK IS a small community, a village really, in the heart of Iowa's Grant Wood country. Forty-one of its households lie in a hollow along the stream that gives it its name, so that in summer only the white steeple of the Methodist church and the dense foliage of cottonwood and maple trees can be seen from the paved road as you drive toward it. Other families among the 369 people who call Crow Creek home live on scattered farms within a radius of a mile or so. This is densely settled country, all rolling hills, woods, limestone bluffs, and cornfields, with deep, fertile, well-drained soils and long, hot, humid summers. The road through Crow Creek that heads north a couple of miles to the meandering Wapsipinicon River has never been paved. Just half the size of Prairie, and in a far more picturesque setting, Crow Creek's gravel road, old trees, white frame houses with open porches, and big green lawns give it an old-fashioned look.

Looks in villages deceive. The meanest village in Egypt or India—all mud-walled lanes, flies, dust, and squalor at first glance—can turn out to be a universe of human relations, with a complex and varied life and culture. Crow Creek is like that. It looks so pastoral and rustic, you'd never guess it is full of urban invaders like the Alvarezes and Novaks, commuters who have moved in from the city. Or that there are so many nostagic old-timers too, like the Mitchells, the Biggses, or old Mrs. Pierce. Or that there are farmers that stayed too small like Adam Pierce, or got too big like Miles West, or farm land in what used to be five farms, like the innovative young Clint Houseman, an outsider who is hardly part of Crow Creek at all.

Most of the time Crow Creek's people go their own way, daily living out their own experience. All are torn, perhaps more than many people, between wanting to exercise freedom of choice and the cultural force of a rural way of life shaped over a very long time. It is a paradox that the sense of removal is greater in Crow Creek, with its dense rural settlement and nearness to a city, than in Prairie, geographically so remote yet somehow more in touch. Crow Creek's farming families, men and women of feeling intelligence, find their cultural survival essentially threatened by the steady urbanization of America in the late twentieth century.

To look at it, this eastern Iowa landscape has not changed all that much since Grant Wood stood at his easel in a hillside cornfield in 1930 to paint Stone City, the next village just a mile or so down the road, which straddles the Wapsipinicon itself. Today you can stand on the same hill Wood did and see the same stone Catholic church to the left, the limestone quarry that cuts deep into the facing hill, the old stone grocery store, a red-painted water tower, farms and windmills, and the road from Crow Creek crossing the bridge and winding like a ribbon up over the hills toward Anamosa and out of view. Wood painted *Stone City* and *American Gothic* one after the other in the same year. Like the Statue of Liberty, the second—a house, a man and woman, and a pitchfork—became an American national icon. Somehow it's all there: our Puritan origin, frontier life, free enterprise, self-reliance, the Protestant work ethic, farming, the family. It's like a legendary and often lampooned—for humor is part of the inheritance too—portrait of our roots.

Four years earlier in Paris, in 1926, Wood had told William L. Shirer, also a local boy—from Cedar Rapids while Wood grew up on a farm near Anamosa—that he had decided after years of studying art in Munich and Paris that an artist, any creative artist, ought to deal with things he knows and all he really knew was home, Iowa. "The farm at Anamosa," he said. "Milking cows. Cedar Rapids. The typical small town, all right. Everything commonplace. Your neighbors, the quiet streets, the clapboard houses, the drab clothes, the dried up lives, the hypocritical talk, the silly boosters, the poverty of . . . damn it, culture . . . and all the rest."

Shirer, in his *Twentieth Century Journey,* quotes Wood as telling him: "I'm going home for good. And I'm going to paint those damn cows and barns and barnyards and cornfields and little red schoolhouses and all those pinched faces and the women in their aprons and the men in their overalls and store suits and the look of a field or a street in the heat of summer or when it's ten below and the snow piled high. . . ."

Shirer himself, like North Dakota's Sevareid, went on to become a CBS war correspondent and the author of *Berlin Diary.* Today he lives in

Massachusetts. Unlike Sevareid, Shirer never idealized the Middle West, maybe because he was a town boy. In *Twentieth Century Journey* he describes his native eastern Iowa as "the heart of the Corn Belt amidst one of the richest farmlands on earth." But he found its people "full of Christian zeal" and "unduly conservative, rigid and complacent, somewhat smug, a little snobbish and determined to clamp down on the rest of us a deadly conformity to their narrow mores, which is one mark of all our small towns. . . ."

To me, Iowa remains something of the alien corn to this day; the North Dakota plains are home. But many summers when I was a boy we went to Crow Creek where, in those days, two of my mother's sisters lived, both married to farmers. At home just about everybody but us seemed to be Scandinavian or German-Russian, first- or second-generation immigrants, children of Grieg and Ibsen. When my father as a country doctor went to see patients in bad winter weather, the old Norwegian who drove his sleigh knew hardly any English. In Iowa, among Methodists and Quakers of English stock, life was gentler than up on the raw, wind-chilled North Dakota prairie, though there was the same gloomy Protestant austerity. You mowed hay, put up bales in the great big barns they still had in those days, helped get the cattle in; or sat in cool dark rooms in big white frame houses with enormous lawns, learning by heart poems by Whittier and Longfellow and listening to an aunt play Chopin and Rachmaninoff. The main event of the week was going to church, where stone-deaf Elsie Brown led the Sunday school, and Harry Newlin, who ran Crow Creek's one business, a produce that bought the farmer's eggs, led the singing. Every service ended with a loud and fervent rendition of "Blessed Be the Tie That Binds."

Two summers, 1932 and 1933—I was still just an infant—Grant Wood held an art colony in Stone City. The artists brought in a lot of old wooden ice wagons, painted them in bright colors and murals, and lived in them. Wood was described by *Time* magazine as the "chief philosopher" of the Midwest school in American realist painting. He wanted to found a grass-roots arts colony like Taos, New Mexico. His experiment lasted just the two summers, but for a time artists were all over the place, even Crow Creek. People still tell about the time one artist was chased off her porch with a broom by old Grandma Pierce after he wanted to sketch her as "a vanishing American type." That was over 50 years ago.

And a lot *has* vanished, even if the houses struggling up the slope from Crow Creek past the church to the schoolhouse and open country and cornfields are better painted and the lawns better tended than ever before. In 1990 Crow Creek was 153 years old. Its first settler, a 21-year-old South Carolinian whose name was indeed Edward Crow, first set foot in it in 1837 when he cleared a patch of land in what was a thick black walnut forest and staked a claim to it. He came, the old records say, with a younger brother and some friends over what the pioneers called Big Woods Road, after crossing the Mississippi on a log raft they had to build themselves. They came in September, and before winter set in, they built log cabins and cut and stored hay for their horses; once the snow fell, they hunted, tanned hides, and traded with the local Indian tribe, which gave Iowa its name. Edward Crow's grave is out in the cemetery on a wooded hill just east of town.

After 6 years Crow Creek had enough settlers for its first schoolhouse, also a log cabin, built in 1843. Carnie Marshall, who arrived with her family in an oxen-pulled covered wagon in 1846, carried six hundred seedlings with her; scattered about the woods along the creek you can find the now-wild progeny of her peach, plum, cherry, pear, and apple trees. The Dubuque Southwestern Railway, with plans to link this frontier with Chicago, came through with a small branch line in 1861 and opened up Crow Creek Station. A depot agent came in, promptly became justice of the peace, and set up a sawmill; it prospered for 10 years until all the choice local walnut trees were cut down.

As in many of the ever-advancing Great Plains settlements, Crow Creek had no church; none got built until the 1880s. Religion in this land of log huts, hand-hewn wooden fences, and coonskin caps came in the form of Methodist circuit preachers, who traveled the forest and farm trails on horseback, sometimes sleeping out in the woods, sometimes in a settler family's cabin. A stump, a block, or a log served as a pulpit; rowdies came drunk to their camp meetings; a pair of fists, a Bible, and a fervency to save souls were all that was needed to expound the word of God.

By 1870, just after the Civil War, from a few cabins huddled together in the walnut trees and cleared patches of land, Crow Creek had already grown into quite a place, with two general stores, three two-story wooden hotels, a gristmill, a saloon, two wagon shops, three blacksmiths, two shoe shops, two egg merchants, a barber shop, pool hall, sawmill, lumberyard, two tailor shops, a dentist, a lawyer, a justice of the peace, and a town constable. By the turn of the century, there was even a skating rink and dance hall. Even so, Crow Creek was outshone by Stone City down the road. Until Portland Cement came along, Stone City's limestone quarry flourished; the little town even had its opera house, where Jenny Lind once came to sing. Down the banks of the Wapsipinicon was the Matsell place, where peacocks strutted on the lawn of a pillared mansion. There lived a former police chief of New York City, who fled when the courts caught up with his cronies in Tammany Hall.

Today most of the old splendor of Stone City lies about in overgrown limestone ruins that seem as old as those left by the Romans. In Crow Creek itself, there is nothing to show it ever had more than its little post office and one general store with a gas pump out front. This is on Pleasant Street, so named by the Civic League a generation ago; it goes to Church Street at one end and School Street at the other. Both peter out in a couple of blocks, and that is just about the whole town.

On weekdays, except for a few retired people, Crow Creek is deserted. Perhaps as many as three-fourths of the wives in its forty-one households work, and half the town is employed by the Collins Division of Rockwell International, in Cedar Rapids just 25 miles away. With 8,000 workers in seven towns and cities, Rockwell is, after John Deere, Iowa's biggest employer. Its Collins works makes super-high-tech radio equipment, aviation electronics, and the instrument panels of many of the world's jet aircraft. Collins avionics are used on American spacecraft. It is quite possible some of Edward Crow's, and his fellow pioneer settlers', descendants assembled some tiny electronic components that helped put man on the moon.

So in a century and a half, the economic basis of this small Iowa community has pretty much gone from agriculture, and initially primitive slash-and-burn cultivation at that, to space-age avionics. This is a radical jump, even in that much time, for it spans most of mankind's technological advance. Crow Creek's own people say they have gone from a farming community to a "bedroom community," a place where people who go to work in the city can live in a rural setting.

Actually, Crow Creek's 369 people include retired people and urban commuters among the forty-one households in town, plus, as an unincorporated community, urban commuters who live on farms in the surrounding countryside but do not farm themselves, urban commuters who farm part-time, and full-time farmers. In contrast, the 761 people of Prairie, a proper town and county seat, practically all live in the town itself. As a county seat, Prairie draws on—if the Hawk County Fair attendance is any indication—a larger and looser community of about 2,000 people, some of whom live on farms up to 15, 20 miles away. Crow Creek is a much tighter little community. The three farmers we will meet, who farm virtually all the land immediately around Crow Creek, land which used to be farmed by fifteen or twenty families, all live within a quarter mile of Crow Creek. A fourth farmer, interviewed because, like Alfred Zuckerman, he is so traditional, is 3 miles away but is already in another community altogether, the Quaker settlement of Whittier.

Prairie is physically spacious, something we see reflected in the mentality of its people. Crow Creek, in contrast, is much more ingrown, narrower, and so are its people. Prairie cannot survive without farming; all it has is its agricultural base. Crow Creek, which exemplifies the other main American rural pattern, is close enough to a city so it can economically survive. It is Crow Creek's old farm-based rural culture that faces extinction.

Ask any of Crow Creek's old-timers today what has changed most in their lifetimes, and they all mention how many farms there used to be. "In the old days," says Sarah Pierce, a farmer's widow and retired postmaster of 86, "there were a lot more farms. A lot on the road to Stone City were just 5 to 15 acres. The bigger farms were 80 on up. There were so many farmhouses. Along the farm roads now, there'll just be an acre with a house and that's all.

"Our house out at the farm was white frame. A six-room house with a woodshed attached to it. We had windbreaks. West of the barn. They were pine trees. I imagine they were planted in the 1870s. Oh, yes, there was an old barn. About 30 by 40, an old frame barn built years and years ago. They used to paint barns red. It was cheap. They used linseed oil and Venetian Red. And they would usually trim it in white. All the roads were dirt and all the barns were red. I expect nine out of ten barns are white now. It makes them look larger. Now so many barns are not kept up. A lot of them just go to pieces. Derelict. I just hate that. I can hardly stand that." Mrs. Pierce came to Crow Creek when she married in 1924. She recalls there were three general stores left then, groceries and dry goods and one also served as the post office. "Elsie Brown. She always kept, oh, fancy dishes. If somebody was going to get married and set up housekeeping, Elsie'd order them a set of pretty dishes, utilitarian but pretty too."

Ed Biggs, 79, who retired as the school bus driver and custodian in 1975 after 25 years—his wife, Elvira, was the school cook—lives in a house his grandparents built in 1889. "Grandma ran a boarding house for the peddlers who went through town. They'd come down on the train from Cedar Rapids, and they might be in town here a day or two. They'd rent a horse and buggy at the livery stable, and they'd go around selling their wares." Ed says Crow Creek probably looks better today, now that you don't have all the soot from coal and wood fires.

"The coal yard was right down here. There was no insulation in those days, just siding on the outside and lath and plaster on the inside. The wind would blow right around the windows. Blow your hat off when it was cold in winter. Just like a barn. Then they only painted 'em once every 20 years. Now today, you wouldn't find any paint that's gonna last 20 years. Many houses are covered with aluminum nowadays; it looks like wood, but it's siding aluminum." It was probably fortunate they didn't paint oftener. Mrs. Pierce told me in those days they used white lead and linseed oil. "Children licking the paint got so much lead. They just found that out in the last 20, 30 years. Good Land, some of those children got real sick, almost died from it, and we didn't know what it was. You don't use lead in paint much anymore."

Ed Biggs, a lean, spry, unusually active old man in faded overalls and a seed cap, says he and Elvira, who is now in poor health and also grew up in Crow Creek, after their marriage, went out and lived in Colorado for 12 years; "We just wanted to see the West." They came home when their parents started to age. "Her folks was old, my folks was old, time to come back here and look after them. And we buried them old. We're still here. We both worked at the school: I drove the bus and done the janitor work for 25 years and she was hot lunch."

As someone who grew up in Crow Creek, Ed's memory of it as a child goes back to World War I. I asked him how its way of life had changed. Ed: "Good Lord, talk about changes. People used to walk a lot more. I can remember when I was a kid about 12 years old. On Saturday all the men walked into Anamosa and got their groceries. That's 7 miles each way. If they got a ride in a buggy or a horse and wagon, that's fine. But you know them days it was clay roads. When it was muddy, you didn't go nowhere in a car. You walked or you rode a horse and buggy.

"Or took the train. My two sisters hopped on the train down here in the morning, went to Anamosa to high school, and hopped on the train coming back at night. Time after time you'd have three steamers pulling the freight cars; there'd be a hundred cars behind those steamers." He grinned merrily. "It used to tickle me. Once in a while one of those trains would go bangety-bang. One guy, the engineer, would get to pushing a little too hard, they'd slip, and the cars would rattle on the tracks. The other two engineers, they'd toot-toot on something, and that'd give the guy hell for trying to be in such a hurry. And Grandpa would say, 'They're talking to one another with that whistle.' " He gave a deep sigh. "The jitney quit in fifty. They pulled the tracks out of here just 5, 6 years ago."

His father, Ed recalls, farmed north of Crow Creek, and when he was a teenager, he and his father milked twenty cows every morning and night. A family could make a good living on 80 acres, with eight or ten cows, fifty or so hogs, and hay and grain and corn. Ed says about the only time people planted oats was when they wanted a nurse crop to start seeding with alfalfa or clover or timothy. That way you got oats one year and the next you got hay. "You fed the horses oats and hay and corn. You get some heat in corn. You got bulk in hay and oats, but you got heat in corn. You know, years ago if corn made 50 bushels an acre, that was a hell of a crop. Now it makes 300. Not all corn, but some of the crops nowadays, ye gods, it's terrific." He shook his head. "It's all hybrid nowadays. It used to be open-pollinated. Yeah, there was a lot of corn that didn't make 25 bushels—hilly, poor ground. Them days, you'd pick 2 or 3 acres and get a wagon box full. And who milks cows nowadays? Dairies. A normal farmer ain't even got a cow on the place, unless he's feeding beef cattle or something like that."

He scratched his head. People were a lot poorer in those days, it seems to him. "I can remember when people used to take the preacher a dressed chicken or something like that once in a while, when they didn't have any money. Keep him eating, 'cause in little places like Crow Creek they didn't eat too good and neither did anybody else."

I asked him, with fewer farms, were people more removed from nature? He screwed up his eyes and thought a little. "Oh, God," he finally said, "I don't know how to answer that. Kids years ago raised on a farm, boys and girls both, were well aware of having little calves and little pigs and chickens. Didn't think nothing of it. Of course many people nowadays have no idea what the hell takes place. One time when I was this little kid, a cow was having a calf and she was having a hell of a time. It got turned around someway or other. My dad got a bucket of soapy water, and he run his sleeve up and got his arm all soaped up. He went in that cow and turned that calf around, and the calf come out and they both lived happy ever after, I guess. He knew what to do. I don't think I'd know what to do today."

Ed also thinks the old-fashioned schools, before they consolidated several together and got so big, were probably better. "Well, six, eight, ten in a class, if the teachers couldn't get it into your head, they poked it into your head. I mean they put in time with you. They was talking with each one of them all the time. And kids in such a small class can't raise as much hell because they're right in front of the teacher. Nowadays instead of getting some knowledge out of a book, they're getting a lot of nonsense.

"At least some of us older people say that's nonsense. You can get that computer and visual aids and seeing chickens hatch and stuff later in life." I've heard Oxford professors say much the same thing: the old-fashioned system, when you drilled memory, was the best education. As I mentioned earlier, I feel strongly children ought to learn by heart dates, places, history, names, mathematical tables, and above all, poems.

At the county fair in Prairie, we heard the emcee, Rube Liebman, ask the grandstand crowd in 1929 how many had heard of a radio 5 years before, seen a plane 10 years before, or owned a car 15 years before. Of the three, the car had the biggest impact on Crow Creek. Ed Biggs says, "When I was young, it was quite a ways to go to town. Now it's 20 minutes to Cedar Rapids, 10 minutes to Anamosa, 15 minutes to Mount Vernon. A store here can't compete. Too much cut-rate. Too many Piggly-Wiggly stores, Me Too stores, supermarkets. You can't compete with that. If a loaf of bread costs 75 cents in Anamosa and 90 cents here, they'll wait till they go to Anamosa to get a loaf of bread. Just like gasoline. If they're selling gasoline in Anamosa for 90 cents a gallon and they have to pay $1 a gallon down here, people will wait until they go to Anamosa to buy gas." This isn't quite true. In 1989, when a local couple who had lived in Florida came home and reopened Crow Creek's grocery store and gas pump, a lot of people were so relieved to see a sign of life in town again, it turned out they were willing to pay a little more for gas and bread if it kept the store open.

The big change in Crow Creek came in World War II. "The war got everybody to working in factories around here," says Ed. In Cedar Rapids a small radio-manufacturing company, started in 1931 by a local man, Arthur Collins, got off to a flying start when it was chosen by Admiral Richard Byrd to provide the radio equipment for his second, 1933, South Pole expedition; Byrd stayed alone at an advance base camp 123 miles from the South Pole for several winter months, making observations and cut off from the world except for his trusty Collins radio. In 1936 Collins entered the new aviation electronics business, making radios for the first airlines (Pan American started up in 1927, TWA in 1930, United in 1931, American in 1936, Eastern in 1938). The big expansion came when

Collins Radio outfitted U.S. forces with military communications equipment during World War II, with its "Roger!" and walkie-talkies and SOS-signalling crank-generated portable radios, or "coffee grinders." Iowans were first worried when Arthur Collins sold out to Rockwell International in the 1970s, though the big multinational corporation has steadily expanded its plants in Cedar Rapids.

If Crow Creek's people began in the 1940s to work in factory jobs in the city, it was just a matter of time before factory workers in Cedar Rapids would start moving to Crow Creek to live. Alice and Jonas Mitchell, whom we'll meet in a minute, were the first to come, buying an old house across from the Methodist church in 1947.

Today commuters are in the vast majority. Ed Biggs says Crow Creek is getting to be what he calls "a community of outsiders." Ed: "It pretty much is. The old people died off and new people come in from the city to buy cheaper property. You know Alice and Jonas Mitchell and Sarah Pierce and ourselves and Louise Fisher and her sister—we're about the only old, original people here. Oh, and the Wests and Pierces and a few others at the church, if you count the farmers. Why, Jane Pierce Scot has been pushing that organ at the church ever since she was that high. She must have been playing that organ for 25, 30 years."

He goes on, "Crow Creek would be dead if it didn't have the Methodist church. And the Civic League." Since Crow Creek has always stayed too small to have a local government, the Civic League, a group of women, mainly recruited from the Methodist church, meet to run whatever needs it and to raise money for road repair, street signs, and streetlights. I know years ago, when my mother went to meetings for a time, she got the Civic League to name the streets, herself coming up with Pleasant Street for the main drag, a two-block-long stretch of gravel, now ending in the grocery store and gas pump. "These women," says Ed Biggs, "they get a little enjoyment out of trying to keep the streetlights burning and various things. Clean up the town every once in a while, you know."

Louise Fisher, whom Ed mentioned, is still one of the stalwarts of the Civic League, as well as the church. Like him, she married and moved away, only to come home with several children after her divorce in the 1940s. Nowadays, with half of American marriages ending in divorce, nobody would think a thing of it. But in those days it was a scandal. I was 8 or 9 at the time and used to think of Louise as looking sad and outcast, like Hester Prynne in *The Scarlet Letter*.

"I raised all my children in Crow Creek, and I never could have done it, not the years alone, if I hadn't been in this town," Louise, now in her 70s, a tall, spare, careworn-looking woman, told me. "Because everybody accepted those children. They were part of Crow Creek even if their father had gone off." She remembers moving in from a farm in 1921. "The telephone exchange was next door. Our ring was a short and a long. Everybody listened in on everybody else. Upstairs, over the store, they used to have dances. I was never allowed to go, even to just stand around."

I asked Louise about the Civic League and its members, a once formidable platoon of ladies my aunts used to entertain. "A lot of them have died off, to be honest," Louise said. "Younger women can't come because they're working somewhere. We have our meetings at night, but even so, they don't come. There's too many outside things now. And there's TV. Cedar Rapids is near. I miss the old parent-teacher meetings at school, too."

So life in Crow Creek has changed, I asked. "It *is* different. Definitely. I think losing the high school, when they consolidated Crow Creek, Martelle, and Anamosa into a single school district and began taking our kids to Anamosa by bus, that took something away. Before, all the kids and parents would go to school once a month. Cookies and coffee afterwards. It was a big social time. And you had basketball games. And baseball. Everybody came. Now, if we lost the church, we wouldn't have anything. I mean, now everybody goes their separate ways."

Shortly before my last visit, Louise Fisher's son Hank, an unmarried son who lives with her, won a $25,000 lottery. "It was the worst thing that ever happened to him," his mother told me, with a laugh. After taxes it came to $19,000, and Hank bought a $16,000 Ford Bronco. With the rest of the money he threw a big beer party, inviting everybody in town.

Another Civic League survivor is Martha Fletcher, a thin, elderly farmer's widow in her 70s with an erect, elegant look, who now lives in a condominium in the next community to the west, Springville. When I asked her how Civic League was doing, she, too, said few younger women took part. "You're not going to get a mother who's working to supplement an income to take much part in Civic League. Sad as it may seem. And you aren't going to get her to give up her job and come home to look after her children and do things for the community. She'll probably get criticized for not doing it. But things are changing. You can't hold on to what was."

With several grown children off living in cities, Mrs. Fletcher lives alone. She feels many of the smallest rural communities in Iowa may survive, but without their businesses. "Stone City's store folded up," she told me. "You look at Martelle. What's there now? Can you even buy a loaf of bread? They go. And it creeps up to the next ones larger. It's just creeping up. You can't go back. You can't go back. You can't live life in reverse. Who wants to go back to washboards? And carrying water in pails?"

Martha Fletcher's husband, who would now be in his early 80s, had died by my last visit to Crow Creek. Tall and lanky, Harold Fletcher had been one of the main farmers south of town and a pillar of the Methodist church; the Fletchers and their children rarely missed a Sunday. Once, on an earlier visit, I asked him about the old days in Crow Creek. Harold Fletcher: "It was a way of life. In the 1930s and forties when all my brothers were farming, we burned wood, had kerosene lamps and ice-boxes, no refrigerator. Oh, living here as a kid. All the neighbors looked after you. And they didn't hesitate to say, if you got out of line a little bit, 'Your dad or mom wouldn't like it.' Neighbors saw the kids doing something, they hollered at them. Now, oh, boy, it's all: 'I don't want to be involved.' "

A big change, he thought, was the way everybody used to keep their barns painted and repaired. Now hardly anybody does. "Now those big old barns are so obsolete," he said. "They can't use them to store hay, with all the labor it would take and all. Why should they put the money into repair? Barns like that were built for the old loose hay we used to put up in small bales. Now that they have these great big bales, many barns are standing derelict. Of all us Fletchers, just a cousin's son is still farming. And he's got a job in Cedar Rapids and just raises sheep on the side."

Even I can remember the days when hardly anybody in Crow Creek had indoor plumbing. You went out to wooden outhouses with old Sears-Roebuck catalogues and spiderwebs. Water came from wells you had to draw with hand-cranked pumps. You filled a bucket and if you were thirsty, as in Iowa summers you practically always were, you drank out of a dipper, which made the cool water taste of tin. In July 1944, when I was 13 and spending a summer there, Crow Creek had its worst flood in memory. I wrote home to my family in North Dakota about it: "The sky was black and torrents of rain came down. Streams of water were gushing across the road from the pond to Everett Brown's pasture. The road was practically being washed away. . . .

"When we got downtown we saw the depot had floated away. The creek had rose up to Mrs. Benton's house and reached the windows of Mrs. Petty's house. The bridge across the creek was gone. Scores of trees were broke and bent. Harry Newlin, plus helpers, was carrying egg crates to higher ground. Two floated away, heading toward Stone City. Across the creek, Mr. Potter and family were trying vainly to start their old jalopy. As a last resort to keep it from floating away they chained it and weighted it until the rising water chased them up to their house.

"The creek's water was black with topsoil from neighboring farms. Many crops have been ruined. Mother Nature has shown her powers and is not looked upon as meek or gentle. . . ." Elsewhere I mention we said "The Twenty-third Psalm" and sang "This Is My Father's World" at Sunday school and that after the flood waters went down, while wading in the creek, I stepped on "a poisonous green, brown and white snake." But there is disappointingly little from this eyewitness to tell us what life in Crow Creek was like.

Alice and Jonas Mitchell, I found, were generally seen as fellow old-timers by others their age. Unlike the other people who moved in from the city, they are not regarded as outsiders. Then I found out what made their difference; both of them had grown up on farms. As luck would have it, it was in the same relatively small area in northeastern Iowa where my grandfather, Rev. Williams, had preached off and on (1890–1920), my mother taught school (1905–10), and Norman Borlaug was born (1914) and raised. Jonas Mitchell was born in 1908 and, like Alice, has exceptionally good recall.

The little communities of Jericho and Saude—Borlaug's home—had many old Norwegians, Jonas Mitchell, his eyes still bright and lively at 82, recalls, and they talked Norwegian all the time. From Saude to another neighboring town, Protivin, everybody was Czech. Over by Lawler, they were Irish Catholic, and west of New Hampton, German Catholic. "Those Germans didn't want anybody else in their town," Jonas recalls. "The Norwegians were the friendliest. The Irish were fun, but you had to watch out; they were always fighting. I mean when I was a kid and went to dances. When the Irish started to fight, you wanted to leave."

Along with this ethnic mix—Dr. Borlaug has already told us how Dvořák was inspired to write his *New World Symphony* here—both the Mitchells remember Jewish peddlers. "Most of them had one horse and a wagon and they had brooms and shoestrings and pencils, all those goods," Alice Mitchell remembers. "One old peddler, he used to measure a yard from the end of his nose. He sold cloth. He bought old papers and rags too. In our family, we kids could have the money from the rags. So, oh, did we ever save rags! And he'd put them in a gunnysack and he'd weigh them. On Friday night from sundown to sunrise on Sunday, they wouldn't work. My mother, when she was a girl, she would go and milk their cows and feed their chickens and gather eggs, because they didn't work on the Sabbath. Up at the New Hampton cemetery, their tombstones always faced the opposite way. In New Hampton the Irish Catholics and the Polish Catholics buried their dead in separate places too. They still do."

Farmland was extremely rocky and Jonas Mitchell's father would dynamite it, buying the dynamite by the 50-pound boxfuls. He raised corn and oats and barley, and if the land was too wet, he put it in buckwheat. Jonas, now aged and bald headed and ailing, can remember hauling molasses over to Saude, Dr. Borlaug's rural trading crossroads, where it was cooked into syrup. All the farms had cows, calves, hogs, sheep, horses, geese, ducks. "There was chickens running all over the place. We even tried to raise turkeys." Jonas said a heart problem keeps him at home these days.

Both he and Alice went to one-room schoolhouses like one not far away where my mother taught. Jonas remembers great big boys, 16 or 17, who only came in the wintertime, the 12 weeks required by law, as mentioned, after 1979. There was a creek near the schoolhouse and in the winter every boy got a stick and skates and played what they called shinny, a form of hockey. "When spring come," Jonas says, "as soon as the creek melted and gophers come out, why we got buckets and we'd catch the gophers. Chased the girls with them." He chuckles.

"I think those one-room schools were good," says Alice Mitchell, a gentle, sweet-looking white-haired woman. "There was a lot of grammar and reading and math. And penmanship was important. Every month we had to memorize two poems from our reader. I did that from 1919 until I graduated from eighth grade in 1928. My mother had been a teacher and she was great on poetry too."

The only traveling show Alice can remember was *Uncle Tom's Cabin*. Her first movie was *The Singing Fool*. Both Mitchells recall threshing with nostalgia. Jonas worked as a bundle hauler and was always in a horse race with somebody. Alice remembers how, as a little girl, she went out to blow the whistle on the threshing machine. "That was more fun. You pulled it so all the farmers knew it was time to hitch up their horses.

"I often think about cooking for threshers. You always had a white tablecloth. You stretched your table way out. We'd have to pump all the water, we kids would. And we had big washtubs and we'd set them out in the sun and have a wash pan there, and the soap and towels and a mirror.

The men would have washed off the worst before they came in. For supper you had to turn the tablecloth over because it was so dirty from dinner. It wasn't so easy to wash in those days. . . . I can remember dressing chickens with the windmill pumping cold water on my hands. It was important to get them cold. Mother always served chicken to threshers. Once threshing was over and Dad had a free day, Mother and I would do the chores so he could get up real early and go fishing. We'd take care of the cows. Of course, we didn't have to do much for the horses. It would be summertime and they would just be turned out to pasture." She smiled happily at the memory of it. Of course the life could be hard too. Like Jonas, Alice can remember her parents, if they got a little extra time, hitching up the team to a wagon and going out to their fields to pick up rocks, trying to improve the land. She sighed. "Our folks both lost their farms in the Depression. To this day we can't go out and spend too much money without really squirming. Because we went through that with our parents when it was so hard. Always, we've watched our money. It's really affected our lives."

As with so many Depression victims, World War II saved the Mitchells. Jonas got a job with Arthur Collins, and after a few years they decided to move out to Crow Creek. "Jonas was Crow Creek's first commuter," Alice recalls. "It was just after the war, Jonas was working at Collins's, and you couldn't buy anything in Cedar Rapids. I thought it would be great because we had three boys and it seemed a wonderful place to raise them. My daughter was just going into the fifth grade and she says it was the biggest letdown she'd ever had. I guess I wasn't thinking so much about her." She says their friends thought Jonas was crazy to drive 25 miles to work. Also, in Cedar Rapids the Mitchells lived in a modern house; Crow Creek in 1947, as was said, still lacked indoor plumbing; some people used outhouses as late as 1957.

Ed Biggs said people went a long time without painting their houses. Alice Mitchell remembers it well: "The white houses got very weathered looking. And there were a lot of barns. They must have been for horses. They were never painted. And so many outhouses. Mostly needing paint.

"Everybody raised chickens. The man we bought our place from, he owned the house next door. And he had this big old barn. He was a trucker. Very often if he saw a good bargain at a sale—a calf or some pigs that weren't getting as high a price as he felt they should—he'd buy them and bring them over and put them in his barn. And they'd run in his lot. There was this open ditch that ran in the back. It was an awful looking place."

Alice says Crow Creek, when they came, was all little bogs out in back of the houses. "It was that wet. It was really an open sewer that ran right down the middle of town. Full of sap leaves. And you don't have dumpy looking places anymore." I can remember it well. With all the houses so neat and the yards so trim today, it's amazing Crow Creek and all the other little Midwestern towns looked so different 40, 50 years ago.

When the Mitchells built their present ranch-style house, they tore down the old barn and some big old sapling trees had to be chopped out. Steve, their oldest boy, told them, "If you'll mark the corners, I'll dig the basement." They had a small Ford tractor and a bucket and he did it in a single day. Steve now works for Rockwell; another of the Mitchell boys is a farmer, raising hogs, and a third has his own construction company in Anamosa. I asked Jonas why everybody keeps their lawns so neat nowadays.

"I suppose somebody moved to town and started mowing their lawn and somebody else thought that looked good and didn't want anybody to get ahead of them so they started mowing too. If you didn't mow, they'd all talk about you. We got one over here, straight across right in back, beside Ed. Now that guy, he doesn't mow. He's got a beard too. They call him Whiskers. And the general idea is he's too lazy. He started up some workshops on self-esteem up in Cedar Rapids. Tried to get people here to join them. He's supposed to be a counselor. He needs somebody counseling *him*." He burst into a deep hoarse laugh, which ended in a fit of coughing.

One thing never seems to change in small communities. "What will people say?"—fear of neighbors' censure—is the most potent force in holding things together, whether in Crow Creek or Ghungrali in the Punjab, or Pilangsari in Java. The last time I saw Jonas Mitchell, I asked what happened to Whiskers. He'd taken a job at the reformatory in Anamosa and was now keeping his lawn as neatly mowed as everybody else.

Alice notices how much younger Crow Creek's people seem today. "When we first moved here," she says, "there were an awful lot of old ladies. Nellie Patterson, Fanny Brown, Elsie Brown, Edna Newlin. . . . When I think of Crow Creek, I think of old ladies living alone. Now I can think of just one house where there's an old lady living alone. My boys quite often hauled water or filled the oil burner for those old ladies. There aren't those kind of people here anymore. It has changed. So many young people with children."

It is curious. People say the same thing in Prairie, making you wonder if there isn't some demographic shift. I remember the old ladies well. Nellie Patterson often waved from her big front porch. Fanny Brown lived in a yellow house set way back from the road and seemed rather grand. One summer she came back from a trip with a bunch of evangelists, real holy rollers who put up a tent in her backyard and held camp meetings. Fred Bailey, a notorious town drunk, went forward one night to be saved. Elsie Brown ran a store and the post office and the Sunday school, even if she couldn't hear a word. Edna Newlin, who played the organ at the Methodist church, was so farsighted she sat stiffly with her head tilted way back and her arms stretched straight out to the keys. They had all been in the cemetery for 30, 40 years now, vivid personalities every one and the hard core of the Civic League. In Prairie a wave of funerals for elderly widows came in the 1980s, but then Prairie was settled a good 50 years after Crow Creek was.

"Nearly everybody in town used to belong to Civic League," Alice recalls. "And we had lots of things where the whole family came. Like when we'd get together in the evenings and have a potluck supper and sing. Somebody would play the organ or the piano. You know, all those old songs. 'I'll Be Seeing Nellie Home,' 'Take Me Out to the Ball Game,' 'Old Black Joe.' And we had community picnics for years and years. Everybody would come. Play ball. Have contests for kids. Now just about all that is left is our Civic League meetings."

Hardly any women worked outside the home in the old days; now hardly anybody stays at home. "Everybody wants a car and a refrigerator and a TV and a freezer and a washer and a dryer and a VCR and a computer and a stereo," says Jonas, scratching his forehead. "And eventually a motor home and a boat and a four-wheel-drive pickup. It's very rare you'll find the mother who says, 'Well, my children are too small. I'm staying home.' "

Actually Harold Fletcher was wrong when he said neighbors no longer watch over Crow Creek's children. The Mitchells do. Jonas, whose heart condition keeps him from being very active, says, "Luckily there are a few older retired people at home. Us. Ed and his wife. Tony Novak, who runs the Little League and has diabetes. He's home all day. Louise Fisher. A couple more on the other end of town down by the creek. Otherwise, after everybody goes to work, it'd be a deserted town of children before and after they go to school."

Alice thinks the Little League is great for Crow Creek. "A few years ago a bunch of 'em like Tony Novak got together and fixed up a real nice ball diamond by the school. And they have a Pee Wee team and a Little League team and a Babe Ruth team, and they're starting one for little girls."

Alice keeps an eye out, during the day, on two boys across the street. "Their parents both work and they're home alone. The older boy gets the younger boy's breakfast. I see them, when the little boy leaves for school, the bigger boy is combing his hair. And they do things they shouldn't sometimes. Jonas and I worry about 'em. Maybe our boys did the same things. But when they get a rope tied up in a tree and tie one up by the feet and pull him up and pull him down, and they get three or four neighbor kids helping them, it really scares us. Quite often at nine thirty or ten at night their folks are out calling them, trying to locate them. He's a carpenter. And the mother works in a doctor's office. All the women in Crow Creek work. This is an empty place in the daytime."

One time, Alice says, they were coming home from taking Jonas to the doctor and saw a number of two-by-fours leaning up against one side of the church. "Stakes were pounded in. I told Jonas, 'I believe they must be going to do some work on the church.' Come to find out, the boys had done it to get up on the church roof. Before their parents came home, it was all pulled down and put away."

Some children, of course, love peril. In Iowa summers when I was that age, we used to walk to Stone City and sometimes even to Anamosa on the railroad tracks. It was exciting, as the tracks ran through pastures, woods, and limestone bluffs and you had to be ready to jump off the longer iron bridges in case a train came. You also had to watch out for escaped convicts from the reformatory. They were always getting out and heading toward Cedar Rapids along the tracks to Crow Creek. Alice says she used to worry about it. "I was scared to death of being that close to the reformatory. And Jonas worked nights. You know how often the convicts would escape down the tracks. We'd go somewhere and come home and there'd be guards with rifles in the road. And they'd have you stop and they'd take a flashlight and look through your car. Now it's been years since anybody escaped. Jonas says they've got it so good over there, they need the walls to keep more from coming in."

The Mitchells are people with a strong rural cultural base who adapted to city life, just as they now are adapting to a community of urban commuters. My last visit to Crow Creek happened to coincide with Halloween. The Sunday afternoon before, the Mitchells' son Steve pulled up his pickup on Pleasant Street with a load of pumpkins from his garden.

This has gotten to be a yearly custom, and kids came running from all over town. "I've got free pumpkins for all the kids that have been good," Steve told them. One boy, just a little squirt, asked earnestly, "How much do you have to pay if you haven't been good?" On Halloween itself, the Mitchells told me to come over just as it got dark. There must have been seventy or eighty ghouls, ghosts, creatures with green heads and feelers from outer space, and assorted other monsters coming to the door in the next hour, all shrilly crying, "Tricks or treats!" Alice and Jonas gave everybody a great handful of candy. "Where do they all come from?" I asked. Alice said many were the children of urban commuters who live in farmhouses around Crow Creek. So much of the land is now farmed by a few big operators, many of the old farmhouses, if they don't get torn down, can be rented by young couples for very little.

IT IS CROW Creek's farmers who, with few exceptions, feel themselves culturally embattled. To them, the Methodist church is the last bastion yet to fall to the urban invaders. They still dominate it, as they did the school before consolidation. If the Mitchells, the Biggses, Louise Fisher, and others who live in town are adapting cheerfully if regretfully to change, those on the farms can bitterly resist it. They feel their way of life is going and nothing is taking its place.

One often sees this in the Third World. In Pilangsari on Java, for instance, Husen's father, when dwarf high-yield rice was introduced, adamantly stuck to old ways. He still borrowed and gave loans without interest; he shared one-sixth of his rice crop with even poorer neighbors, mostly elderly women, who planted it and harvested it. He would rather have died than mortgage his land for credit at a bank in order to buy insecticides and fertilizer. Mutual aid had always had a healthy leveling effect, and the father did not like the way some villagers grew richer and others poorer with the new rice. Husen's old mother was equally opposed to contraception. To the elderly couple, value had always been attached to large numbers of children, all-night shadow-play performances, gamelan orchestras, and religious ceremonies and feasts, things that impoverish but bring joy. Again, let me stress why we should avoid value judgments when it comes to cultural change. I sympathized with the parents, but I also knew Java had to grow more rice and have fewer babies to survive.

One of the most sympathetic Sikh farmers in the Punjabi village was old Pritam Singh, a white-bearded giant who might have stepped out of a painting by Michelangelo. Once cash wages were introduced, Pritam was the only landlord who continued to serve his harvesters tea. When I asked him why, Pritam said, "Because they are poor." But when the old man asked a laborer to help him lift a bale and the man refused, saying he wasn't being paid for it, Pritam was enraged. I found him trembling with anger. "That man used to be like a son to me," he said.

In Crow Creek, Adam Pierce, 59, farms about a quarter mile west of town and lives on the old family place just south of town with his wife, Mary, and the two of their four children still at home. It is a big old Victorian brown-shingled two-story house Adam's grandfather built in the late nineteenth century. The front is well maintained by Mary with white-painted shutters and a well-mown lawn with a swing and flower beds. Then Adam's world takes over. The open back porch is inhabited by a menage of cats, kittens, hens, and a friendly mongrel dog. Behind is a woodshed Adam uses as an office. Back of the house, except for a patch of grass under a clothesline, densely overgrown pasture extends way back to the ruin of what was once the biggest barn around. Now it is a great eyesore; half of its roof has caved in. Around what is left of the derelict barn, so gray and weathered the grain in the siding stands out in relief, and half hidden in brush and grass, are the rusted shapes of 40 years of old cars, tractors, and farm machinery, most of it cannibalized. Nearer the house is a sawmill and Adam's machine shed. Along the driveway the family's four cars are parked by the side, those used by Adam, Mary, their son Tom, who after 2 years at a community college works in an audio shop in a Cedar Rapids shopping mall, and their daughter Judy, who is going to college.

The two oldest daughters are married, one to a dentist in Cedar Rapids, the other to a banker in the small town of Clarence, where he farms part-time. Adam himself, after fighting with the Marines in the Korean War and studying to become an electrical engineer, at last decided to do what he wanted to do. That was come home and farm. Adam and I were born just weeks apart, and during my boyhood summers in Crow Creek we were boon companions, exploring the woods and limestone bluffs along the railroad tracks, catching crawfish in the creek, swimming or fishing in the Wapsipinicon, and on rainy days building fortresses out of hay bales in the enormous barn.

A man of medium height, heavy shouldered from the many years of farm work, Adam has never lost a jauntily youthful swing to his walk and movements. When not out at the farm, he is likely to be found repairing some piece of machinery. One day I talked to him as he worked on an engine. Typically his face was grimy, there was a smudge of oil on one cheek near his nose, a cigarette hung from a corner of his mouth, his hands were black, and there were oil and grease on his shirt. Adam's hair is thick and brown—mine turned gray in Vietnam over 20 years ago—and this and some red patches on his cheeks give him a faint resemblance to Ronald Reagan, something Adam will protest strenuously.

When I ask him about Crow Creek, he screws up his eyes, thinks a little, and says, "The Crow Creek we knew as boys died 20 years ago." I protest and he goes on, "Look at it. Where's your little town? A post office open half a day five-and-a-half days a week. For a long time there was no store. There's a Little League, but that's composed of kids from Crow Creek, Stone City, and God knows where. Hell, half the Civic League lives outside Crow Creek. Crow Creek as a town is nothing. It's a collection of houses. Probably half the church is comprised of farm families living outside the town. Probably more.

"No. People in town aren't interested in Crow Creek as a town any-more. And when Wal-Mart came in, it killed Main Street in Anamosa. What ails Crow Creek or Anamosa is just part of the evolution that's been going on here ever since the second world war. When women started to work. That's the big change.

"It may be pride and it may be smug, but you look at my kids. They had a mother who stayed home and looked after them. And then you look at some of these kids. Or just read the papers. All the trouble kids are getting into now. Rape. Drugs. Like that. Kids just going completely unglued. Look how many kids commit crimes. Look how many kids commit sui-cide. Look at the schools they're trying to take over and run as day care centers, for Christ's sake. To provide breakfasts for kids 'cause they don't get them at home. You've had a complete evolution taking place and we're paying for it now."

The Mitchells seem downright cheerful compared with Adam, and he says the worst is still to come: "This isn't anything compared to what it will be. I don't care if you shout 'women's lib' at me or not. But in the family there should be somebody—I don't care which one of them it is, it usually works best if it's the mother—but one of them should be there for the kids. People won't drop dead if they don't have a new car every year. When you have children, you've taken on a responsibility. You've taken on some-thing. Look at those cats playing on the back porch there. The average tabby cat will watch over her kittens more protectively than the average working woman. It's something I feel very strongly about."

Later I was to tie the disintegration of family life in America directly to the decline of farms and small towns. That day Adam seemed to link them, too, but not explicitly. To show me how much had changed, he began taking all the nearest farms, one by one. "You can look around the country-side and you can see how many people are gone. That's right." He pointed down the nearest road. "You start right off. Lafe Warner. Buck's son. Less than 40 and he went bust twice. Borrowed money for machinery and for crops. Well, one year it was pretty dry and he went broke. Buck Warner's oldest boy graduated from college. He could find no work around here, so he's in California. Broadcast marketing or something like that. Buck Warner's little brother, Lester, he's been out in California for some years. A musician, I think. A sister is out in California too, I believe.

"Then there was old Mrs. Hook. Right next door. Had one child, Mary Jane. She married George Bagley. They had two children. Judith's in Colorado. Clem died. Heart attack or blood clot. Now Clem farmed around 40 to 50 acres. He made a little money raising cows. But he had to provide for one person, Clem Bagley. He did it on shares with his mother. He didn't have a family. Never married. He didn't have very expensive wants. And it didn't take much money to live on. Now Clem's gone out to the cemetery. But if he was alive today, he'd need a lot more money. Just your basic things, the increase in the cost of everything. The insurance on your car. The insurance on your health. The insurance on your house. Your liability insurance. Your taxes."

Adam goes on quite a bit about taxes. He says they are the second or third biggest input when it comes to a farmer raising an ear of corn. "People don't wake up to it. Taxes run just under $20 an acre. On that dirt out at the farm. Seed costs about $17 an acre, sometimes a little less. Herbicide about $15, $16. The big cost is fertilizer, which probably runs $50, $60 an acre. Next, on a par—it's a horse race—are seed, herbicide, and taxes. It depends what it's in, but I'm running about 140 bushels to the acre this year. Then you've got the cost of your machinery and stuff like that. It will vary from county to county and the quality of the land. But that's for bare land. No improvements at all. And your taxes are running just shy of $20."

This has spelled an end to the poor, marginal full-time farmer. "You just don't find Clem Bagley types farming anymore. You want to remember that what a person can live on comfortably today is probably three or four times what it was 20 years ago. Now I handled my Uncle Abner's estate. He had less than $10,000 in savings. Why, he couldn't live 5 minutes in today's world. This is something you never read about. But just the sheer cost of living's gone up tremendously. Abner and Bess had 15 acres. The place they lived on. They owned it. You know, at the bend in the road above the Wapsi. Now what did they lack? Running water, that was the only thing they lacked. They had television. They had lights. They got gas. They were paying taxes. That's right. But it didn't cost much. They went out quite a bit. They went to your church suppers. He always had a pretty good car. No, they were doing pretty well right up to the last. He got some Social Security. Raised some ginseng down by the Wapsipinicon. He tended the cemetery. It made him a little bit of money, not a great deal. Abner lived to be 77 years old, he did pretty good. And Bess lived to be about the same age. What would Abner and Bess be doing if they were around today? Like Clem Bagley, they'd all be starving to death."

Adam says you can tell how many people are feeling the pinch by all the woodpiles around Crow Creek's houses. It is to cut down on the sheer cost of living. He cuts his own wood, using a chain saw. Oak, walnut. What is valuable, he saws off for lumber. What is crooked or dead gets sawed up to burn. He goes out to gather wood in the winter about once a month.

Just medical insurance for himself, Mary, and the two kids still at home, he says, cost him close to $3,000 in 1990—"and I got the crummiest plan there is." What worries Adam is that so many jobs nowadays seem to be low-paid, part-time service jobs. In the 1960s, like Andy Rynning in North Dakota is doing today, Adam played guitar on weekends in Kenny Hofer's band, a popular country music band in eastern Iowa at the time. Hofer himself worked during the week at Wilson's meat-packing plant in Cedar Rapids; he started at $6 an hour. That was over 20 years ago, but IBP—Iowa's present-day big meat packer—is paying, Adam says, just a starting wage of $5.

Adam: "There's service jobs and there's a great deal of it, at $5, $6, $7 an hour. Just think what that leaves a person to live on after taxes. Collins has got big and high-tech, they've got space, planes, power tools; hell, it's a big conglomerate. But they're part of Rockwell. And just the minute they get into trouble, Rockwell will dump them. That's the way it is with any of your big firms."

When you look at who is farming what around Crow Creek, it soon becomes evident virtually all the land immediately around the town is now being farmed by two men, Miles West and Clint Houseman. Miles West,

65, was born in Crow Creek and, like Adam himself, is a stalwart of the Methodist church. He owns or rents close to 2,000 acres, a huge farm for Iowa. Clint Houseman, 39, a young university-educated outsider, farms about 700 acres, what used to be five family farms, all of it rented.

Adam admires Houseman's gumption. "Clint's a real go-getter. He's hard working. He's a young guy. He likes, he *loves*, to farm. He went to Iowa State. He's making great headway as far as not tying up a fortune in big machinery. And in getting himself financially sound. I'm not saying he doesn't owe some money and I wouldn't know how much. I do know he's making progress.

"It's a sensible way to farm. He bought a new corn planter a few years ago. He has a big John Deere tractor that's maybe 12, 15 years old. He bought another tractor 3 years ago. But he's not buying a lot of expensive machinery. He uses his head and he plans things out. He has quite a good combine. He put some time and a little money in it and ended up with a good machine. Even with a brand new machine you can spend a lot of time fixing it or sending it to town. I don't think Clint will go on fixing stuff as long as I will."

Adam's own farming operation is much smaller than West's or Houseman's. Farmers in Crow Creek, Clint Houseman excepted, are pretty closemouthed about their farms, unlike farmers in Prairie. West, as we will see, for good reason. As far as I can figure out, Adam, most of his life as a farmer, sharecropped 220 acres owned by an uncle. He inherited half of this farm when the uncle died in the 1980s. In 1989 he farmed 145 acres himself and rented out 80 acres more to Clint Houseman, but in 1990 he turned this 80 acres over to his son Tom to farm. Adam's biggest income over the years has not come from farming but from running a mail-order business in ginseng seedlings and roots; most nights he spends several hours filling orders and doing paperwork, often until 1 or 2 a.m. For a farmer he sleeps late, often until 8 a.m., a habit he picked up from his guitar-playing days.

I ask if any medium-sized farms near Crow Creek survive. "Many mediums?" Adam has to think. "Let's see, Kenny Lukas, he farms 92 acres. He used to be a pretty good-sized farmer. He rented a lot of dirt. And long before a lot of people could see it, he saw there just wasn't any money in it. And he started getting off the dirt. He's a city employee in Springville. And he farms his place at home, 92 acres.

"Bailey down the road here, he farms 240 acres and he runs a tremendous amount of hogs and quite a few cattle. Warners' place, Clint farms that. Ben Fletcher's, I'm not just certain. Martha Fletcher's, Harold's old place, Clint farms. I don't know who farms Powell's now or Shackleford's. Going out this other road toward Whittier, there's me. There's Metzger. He's got 160 acres and he probably runs another 120 over by Springville. He's got two boys and I think they're both out of school. I'm not sure. Next, Garve. He works at Wilson's. He's got 80 there, he farms 100 of Jim Witcomb's and he rents land, I don't know how much. He runs a lot of dirt. Pretty big equipment. And he's got that job at the packing plant.

"Going the other way, Clint's got quite a bit of it south of town. In the 1940s, those were all small farms. Then Collins Radio started taking people. I heard Jonas Mitchell was one of Art Collins's first hundred employees. Way back then. So you had a real good source of income for a lot of people. Crow Creek never went back to being a purely farming center after that. Until then everybody was a farmer or had been a farmer. There wasn't a hell of a lot down there once the city people moved in."

As the small farmers died off, few of their sons took it up. "Look at it this way," says Adam. "If you knew how to do something, but you make a hell of a lot more money doing something else, what are you going to do?"

Adam himself sometimes pays for his independence. Alone of Crow Creek's farmers in 1985–86 and 1986–87, Adam stayed out of the government support program. In 1988 he got back in. It simply cost him too much to stay out. "I usually sell about 5,000 bushels of corn," he explains. "That's $5,000 that the government, by driving down the price of corn, has taken out of my pocket. You pay dearly for standing up for a principle, for trying to stand on your own two feet."

If Adam's family fondly puts up with his ideas and ways—his married daughters still come home with their families practically every weekend— it is Mary Pierce who really holds the family together. A warm-hearted, energetic, hard-working brown-haired woman 8 years younger than Adam, she is up at five thirty every morning to fix Adam's breakfast and lunch before going off to work. In 1988 she did what she said she would never do, and took a job in the city. It is part-time, helping at the cash register and making sandwiches in a delicatessen, part of a big supermarket on the highway into Cedar Rapids. It is 17 miles from home. Mary finds she enjoys getting out now that all four of her children are grown. "The reason one worries more about kids today is that *both* parents are working and nobody's there to supervise the kids," she says. "You see, I don't believe in that. I think if you're going to have kids, you'd better stay home and raise them. Then if you want to go back to work like I did, fine. In Crow Creek a lot of families say the kids are old enough to baby-sit for themselves. I don't think, even in high school, kids should have to come home to an empty house." Polls show two-thirds of American women agree with Mary. But they also show two-thirds of all mothers are in the labor force, twice as many as in 1955. Most disturbing, more than half of mothers with infant children work.

It isn't just materialism, as Adam and Jonas Mitchell seem to think. Many Americans are getting poorer. Average real wages have been in decline ever since the first big oil shock of 1973. In 1985–90 they fell by 5 percent. Part of this is the new competition of the single-world economy. As Calvin Beale says, "If you cannot produce cheaply enough to compete, you've got to reduce your costs and your standard of living or go out of business." The most depressing postindustrial fate is not joblessness. It is the relentless switch from $16-an-hour manufacturing jobs to $4-an-hour service jobs like Mary's. Americans do not like to admit it if they get poorer. As in Crow Creek, they will have smaller families, go into debt, both husband and wife will work, or they'll work longer hours, anything to keep up consumer spending. Even Adam and Mary Pierce are doing this.

A fond, cheerful, constantly busy person with a remarkably agreeable disposition, Mary believes in being neighborly. It worries her how few of the young women among the newcomers to Crow Creek are. "You know, come to meetings all year round. A few old people can't do all this. Alice Mitchell and Louise Fisher and a few others are way up in their 70s, and trying to keep Civic League going. Now Little League, that's completely different. There young people do get involved. About four families got it started. They went around and asked for donations. And Civic League had a big rummage sale and we gave them what we'd collected. The fathers of the kids coach. It's good for Crow Creek. It's fun to go down and see all the people really backing the kids."

Mary finds, now that her own children are grown—she has four grandchildren—that when she goes to a program at the school, she may not know more than a dozen children, whereas once she knew everybody. Lots of them, she finds, now live out on scattered farms around Crow Creek whose land is tilled by somebody like Miles West or Clint Houseman. "I would certainly hate to see Crow Creek lose the school," she says. "Some of the teachers are worried because they think they need at least three hundred to keep it open. I guess they don't have that many, even bringing in kids from way outside. Most people have one or two children. It used to be four or five."

A lot of Crow Creek's young couples, Mary notices, buy everything on credit. She feels they don't know how to save. "Like so many farmers. All of a sudden they go bankrupt. I think we're too much in the opposite direction. Adam won't borrow or go into debt for *anything*. Even to help himself out, he won't do it. He wants to know where he stands from one year to the next."

Adam can't always hold to his principles, when it comes to price supports, as he said. "We had to go back into the corn program because there's no way you can go without it," Mary says, defending him. "It's too bad. I think it's a lot like one of those countries where they can tell you what to do and you have to do it. You don't have your say. No matter how good a farmer you are, there's only a certain amount to be made. You're not your own boss. Somebody's always telling you, and you've got to get rid of your corn. You can't hold it forever."

The crunch for Adam, as with Prairie's Alfred Zuckerman, is that too small a farmer lacks capital to buy new machinery but he can't keep repairing his old tractor and combine indefinitely. Zuckerman wouldn't take credit either. "In farming," says Mary, "you put out so much for machines. Right now we should buy a new combine. But Adam just keeps repairing his old one. There's always something going bad. But you don't see enough money in farming to put it into good machinery. So I don't know. Our boy Tom would like to take over. Adam doesn't want him to do it while he's able to make a good living. You know, better get out and learn his skill first. Maybe later Tom can farm. He could probably farm part-time now. But he's got to have better machinery. Adam works so hard. Most nights he's in his office working on ginseng orders until two o'clock in the morning."

At Adam's insistence, she says, they have no debts at all. "He won't operate that way. A lot of those who are hurting borrowed when interest was 12, 13 percent. Now if they've got any money in savings at all, they're only getting 5, 6 percent. They still have to pay back their loan at the high rate of interest. How are they going to do it? I think the bankers encouraged people to borrow. And now they're selling them out and getting the land back for a little bit or nothing." Land around Crow Creek dropped to as low as $800 an acre in 1987 from a high of $2,000 in the 1970s before edging up again.

Mary expects a rural recovery in the 1990s. "Crow Creek's going to be here forever," she says. "It's so close to Cedar Rapids, it will probably even grow. It wouldn't surprise me at all."

The Pierces face the same father-son continuity problem that Prairie's farmers do. Tom, their son, told me he would love to farm. He figures he would need at least 200 acres to start and would rent more. He doesn't think, as his mother does, that he would need new machinery, but could make do with what Adam has. Judy, the Pierces' youngest daughter, who works hard as a short-order cook in Anamosa's best cafe to help put herself through college in Cedar Rapids, says, "I'd never marry a farmer. Mom does a lot of work but she doesn't get paid for it."

The Pierces have a noisy household full of children and grandchildren most weekends. The whole crowd goes off to Crow Creek's Methodist church, filling several pews. They even take vacations together, everybody piling into a couple of vans to go to a lake resort in Minnesota or, once, to a Jewish wedding in Wisconsin Mary said was like something out of *Fiddler on the Roof.*

I asked Dave Weiss, the son-in-law who farms part-time near Clarence, Iowa, and is a loan officer with a Federal Land Bank, how things were over his way. Dave said the wave of bankruptcies that peaked in 1987 has pretty much died away. Typically, he said, the farmer who went bankrupt had about 450 acres, bought on credit at peak prices of $3,000, $3,500 an acre, and about $50,000 to $100,000 in new machinery, also bought on credit. "We had a farm next to my dad's that even sold for $4,100 an acre in 1985. It was down to $1,700 in 1990. Anybody who buys land at that high price can't pay the interest. They're insolvent. Their liabilities are worth more than their assets and they go bust."

The story on the drought was the same as in Prairie; most farmers suffered worse financially in 1989 when they had already sold the carryover corn in their bins and disaster relief and prices were not as high as in 1988. I asked Dave who was going to survive in farming in Iowa.

"It's going to be the farmer who can produce a bushel of corn for the least amount of money," he said. "Somebody who is a good producer and good financial manager. Also he needs a second occupation. With me it's accounting. You'll always need that." Two-thirds of his high school class of sixteen in Clarence grew up on farms. Ten went to college, Dave to Iowa State. In 1990 two were in Texas, one in Colorado, one a doctor in Cedar Rapids. Two were stuck in unskilled manual labor in Clarence itself. Two, including Dave himself, farm part-time. Just one is a full-time farmer. It is about the same pattern as in Prairie and, one guesses, all over the Middle West.

NOBODY TALKS ABOUT it, but it's been in official notices in the Anamosa paper. In 1989 Miles West, Crow Creek's biggest, proudest, and ostensibly richest farmer, filed for protection from his creditors under Chapter 11 of the Federal Bankruptcy Code. It means West can keep farming while a way is worked out to pay his suppliers and creditors. Evidently, people say, 2 years of drought are what did it.

If you picked a farm couple to pose for a present-day version of Grant Wood's *American Gothic* (Wood's sister Nan and a Cedar Rapids dentist posed for the original), Miles and Betty West would do nicely. West, a handsome, well-preserved man of 65, is better looking and has a lot more hair than the man in the painting, with his pitchfork and bibbed overalls. But he has the same dour, tight-lipped, stern expression. Miles West has a way of standing stiffly, like a soldier saluting, with his dark clean-shaven chin held well in the air. By his well-poised chin, his slight baldness, and handsome features, you feel this is a person of some standing. Betty West, though her brown hair is frizzed by permanents and she doesn't wear the drab dark dress of the nineteenth century, has the same look of pioneer fortitude. One is not surprised to learn the Wests live in that big square red-brick house on the hill next to the cemetery (which Betty West mows these days), upright small-town folks, descendants of the pioneers.

Narrow, cautious, steady, faithful to the Methodist church and, like the Pierces, never missing a Sunday, the Wests exemplify the kind of substantial rural Midwesterner that felt a lot more comfortable with the values of the past. Even in the best of times, nobody works harder, Miles running his huge farm operation and Betty West, now that her two sons are grown and married, raising nearly four thousand chickens. Their tidy house, with its drawn blinds and closed windows, has a museum quality, everything in the parlor and upstairs bedrooms in its place. But the Wests really live in their big, comfortable kitchen where sits a big square table. Over coffee, as Betty frosted a hot chocolate cake she just took out of the oven for us, I asked Miles West how Crow Creek had changed in his lifetime.

West: "When I was young, Dad farmed 115 acres. There were a lot of 80-acre farms at that time. There were a lot of 40-acre farms. We always had about a hundred hens and ten, fifteen cows and a hundred head of hogs, twice a year. And four or five horses to farm with. I was born in 1925, so we're talking early 1930s. I was 6, 7 years old then.

"It was about that time those 40-acre guys started losing out. Not all. When I was a kid, Buck Warner's dad had 40 acres. He always seemed to have money. And there was two little places right down on the edge of the creek, one on the north side coming up and one on the south side, 25 acres and 40. And out there north of town by Adam's, Lester Hook, Clem

Bagley's granddad, he had 40 acres. I don't know what Grandpa Pierce had, but Henry Pierce Junior had about 120 acres of those hills that maybe had 30 acres of farmland on them. And north of our place, out toward Stone City, there were a lot of 10-, 15-, 20-acre places with houses on them. Years ago a lot of men worked in the quarry and farmed some. Yeah, north of here there's plenty of foundations where there were houses, if you know where to look for them.

"The quarry had kind of a dead period in the 1930s, and Adam's dad got a new International truck in 1937. It had a lime-spreader box on it, a real fancy outfit. Before that he just had a truck with a flatbed on it, and they shoved off of that into a pull-type lime spreader. Adam's dad hauled lime part-time and farmed as well. Now they use the quarry for cut stone and facings. Limestone is coming back in."

Like everybody else in Crow Creek, Miles West says the real change came in World War II. Suddenly there were plenty of well-paying jobs in Cedar Rapids, at Collins, Quaker Oats, other mills and factories, and at the meat-packing plants. It spelled the end of the really small farm because earning wages in town was so much easier than trying to wrest a living from 20 or 40 acres. West says even somebody farming 80 acres today has to have a job in town too.

"To farm efficiently today, you've got to have a big operation, 1,000 acres or more," West says. "You've also got guys with 200, 240 acres that have a specialty in something, like cattle or feeder pigs, or Adam's ginseng. A family farm has to be up pretty close to 300 to 400 acres in Iowa to survive. Small farms are gone.

"Small *towns* are gone. The only reason Crow Creek looks healthy now is that it's a bedroom town. That's all any small town is now, is a bedroom town. In that it's good, it keeps the houses full in town. The houses are better maintained than they've ever been. In that it's bad, the influx of some of the people you've got, they're not desirable versus what you had before they came. Put it that way."

What is wrong with the newcomers?

"They don't really have much to do with anybody else. They get up in the morning. They go to work. They buy their groceries in the super-market on the way home. They stop at some place like Wal-Mart and get their other stuff. And they pull into a self-service filling station and fill up with the cheapest gas they can get. They come home and sleep. Their kids are in town, the people like them, and have been from the time they got out of school until the folks get home. And they either got into trouble or some of the neighbors rode herd on them and looked after them. Tony Novak, he's got a diabetes problem and lost both legs and he keeps an eye on a whole bunch of kids that gather in there."

I told West my impression in 1990 was that the farm families were badly hit, but Cedar Rapids seemed to be booming, with a lot of construction, plenty of jobs. It had recovered a lot since 1987–88. He agreed there was now something of a dual economy.

"Most of the new people in Crow Creek have pretty good jobs. At Rockwell-Collins. It's got a lot of government contracts. And so high-tech, I don't know what they do. I know their equipment was the main thing on the first space shuttle. Now, Waterloo's another story. With farm industry and meat packing going out. John Deere cut way back for a while. Waterloo's chock full of unemployed people, a lot of them black."

West listed the occupations in Crow Creek. "Here, besides Rockwell-Collins, there's a couple of welders, a postman, a car salesman, Cecil Peet that retired from Quaker Oats, two or three carpet layers, a carpenter, some of the women are receptionists, one woman's even a truck driver. You got people living in Crow Creek, probably both of them employed in Cedar Rapids, never been laid off. They don't realize the economy is bad. It never has been for them. They pick their paycheck up, they come home. They are doing well. Then you've got a group of people who are self-employed and farmers who are just squeaking by. Or going backward.

"Yeah, there's a new breed in Crow Creek. Jobs in town. Man and wife both work. They like country living. But not a hog smell, they don't want that. They raise hell with whoever is raising hogs. They may have a couple of horses. But they don't like barnyard smells and such. The courts are supposed to side with farmers. According to this 'grandfather' clause. You were there first. But in all cases in Iowa they haven't found that way. They've reversed it a few times and really nailed the hog farmer who was there first. Gave him a big fine. Which isn't right."

The worst side of the urban invasion, Miles West says, is the crime and drugs that come with it, even to Crow Creek. "There's a lot more crime here now than 30 to 40 years ago. A lot of it's never reported. Petty crime. Where stuff just disappears. Nobody will report it because there's nothing going to be done about it anyhow. We got a lot of help around. And you don't lay anything down because it grows legs. And I could lay stuff down here 50 years ago and it'd be there 50 years later. Now, all of a sudden, you lay something down and come back 15 minutes later and it's gone. And nobody saw it. You can't *trust* people."

He gave an example. "David, my son, has a lumber company. And he had this kid working for him in the molding plant. Nice clean kid. Come to find out, he'd been in trouble in California and so he was under a probation officer. Even here it got so his record wasn't so good. If he didn't want to show up, he didn't. Well, one day the truck dispatcher was going to Cedar Rapids and this kid, he needed a ride. And the dispatcher had $400 to deposit that was his and his lady friend's money. He put it in the glove compartment of his pickup. And along the way he got out to go someplace and he left the kid in the pickup. He never thought of the money until he came back out. The kid and the money were gone. Four hundred bucks.

"So they scoured around and couldn't find him. Come to find out, a chain saw was missing too. So we kept after this kid. Never quite caught up with him. How do you trace $400 in cash? Nobody was too interested. Then the Cedar Rapids Police Department called and said, 'Well, we've just had a query on him. A pawnbroker in town just called and asked if there'd been a chain saw stolen. The kid's been in trying to pawn one.' When they picked him up, his wife calls and says, 'We'll pay you the $400 back too if you'll drop charges.'" West's face took on a hard, cold expression.

"We ain't going to drop the charges. He's not getting away with this crap, you know, stealing stuff. He thought he'd make restitution, we'd drop charges, and he's loose. He's already on probation for something he did in California. He's just gone bad."

West thinks the boy may be on drugs or drinking heavily. It turns out drugs are as big a problem in Crow Creek as in Prairie, if not worse. "We've got 'em," West says. "Don't let anybody tell you any different. Anamosa is practically the drug capital of eastern Iowa. It's rampant here. It moved from Iowa City to Anamosa. It started with marijuana, but there's a bunch of hard stuff running around now like coke. Why people use it, I haven't any idea. You just don't walk up and say, 'Why are you taking dope?' When I was on the school board, there was a little floating around. Not bad. It was marijuana. They got that pretty well sorted out. There isn't much going on at the high school.

"The people you hear about aren't high school kids. They're in their mid-20s. They're not kids anymore. In Crow Creek it's not kids. It's adults. They're your 'move-in' element. People who work away from here. We've seen 'em out at the cemetery several times. You see a car coming along the county line road real slow. They never stop. You just see another car coming the other way and an arm shoots out of each car. One time we found a can of marijuana stashed under a tree. They were using the cemetery as a drop. The sheriff come out and got it."

In Crow Creek, as in Prairie, the use of drugs is the most puzzling rural change to me. All rural people, farmers or fishermen, take something. Except in Muslim villages, drinking is common everywhere I've been, whether home brew among Punjab's Sikhs, *cachaca* in Brazil, tequila in Mexico, *tuba* or coconut wine in Filipino villages, beer in Java's Pilangsari or Vietnam. Home brew, the same nasty-tasting, highly potent stuff you got in India, also turned up in Sudan's remote Nuba Mountains and, of all places, in a northern Chinese village about an hour's drive from Beijing. Even in Iran the evening came when, with smirks and secrecy, the doors and windows were bolted and out came a bottle, along with oranges to kill the smell of alcohol on your breath in case, going home, you ran into an ayatollah. Sometimes hashish in cigarettes or a hookah pipe did get passed around in Egypt, Morocco, or Pakistan, but even among Muslim villagers clandestine booze was much more common, I think because alcohol makes people more sociable, not so zonked out as on drugs.

In Anamosa I asked a bartender what the police did about drugs. "Who's going to do anything?" he shrugged. "Somebody might tap you alongside your head with a stick if you did. The cops stay out of sight. The only thing they want to do is issue speed tickets. That's their biggie."

Before leaving the Wests, I want to briefly go into Betty West's poultry operation, since she was one of the few rural women I've come across who ran her own business. "I don't compete with the big poultry companies," she was quick to say. "Not really. I sell to individuals. I'm not trying to go through stores and stuff. As long as I can give them a good product and they know where it's coming from, well, in what I'm doing, I can compete. For broilers, I get week-old baby chicks and keep them for 2 months, April and May. The pullets, or laying chickens, we get by April and they lay in August. They're sold as laying hens."

Mrs. West finds raising chickens is something she can do herself without a big overhead in buildings or hired labor. "A big place, they've got to budget for health insurance, budget for depreciation, budget for this and that. Economies of scale? A big company is not going to get a better percentage of eggs than I get. They're not going to get a better percentage of layers."

Where she can't compete is in selling large amounts of eggs to one of the big local produces. Instead, her regular customers are two grocery stores, two restaurants, a hospital, a care center, and a school. People also come to the door to buy eggs. She sells her pullets as stewing hens. "I get much more per bird. Because, with Miles's help, I'm willing to work harder. I have no extra money in automation. *We're* the automation." Betty West averages about 3,000 Cornish Rock broilers and 850 pullets at any one time.

I tell her I can remember the days when everybody had a few hens. "Nowadays by the time they have to buy feed, it's not worth it," Betty says. "And it's messy." The Wests laugh when I ask about the hens on the Pierces' back porch. "They're Adam's pets," Betty explains. "Mary buys her eggs from me. Adam goes out and talks to them. He grew up with those hens. What happened to that duck, Miles? He had a duck too. And there's cats and kittens and that dog all on that back porch."

The Wests are also amused that Adam stayed out of the corn program for 2 years. "He made a big mistake," Miles says, his face lighting up in a big grin. "There's no way you can stay out of a government program. Just about everybody is in. They subsidize the price, and you take the land out of production. Adam blew it. He cut off his nose to spite his face. He missed the boat completely."

COMPARE THE ATTITUDES of Adam Pierce and Miles West to those of the Mitchells and the other people in town. The two farmers—their wives less so—feel their way of life is being undermined both economically and culturally. West, farming close to 2,000 acres, got too big, took too much credit, and got into trouble. Pierce, farming 225 acres, stayed too small, takes no credit, and is in trouble because he lacks capital to replace the machinery he needs to keep going—or begin to hand down his farm to his son. But it is the cultural change they see around them that makes both men feel so embattled. Adam Pierce says the old Crow Creek, where "everybody was a farmer or had been a farmer," is dead. Except for going to church every Sunday, the core of his traditional culture, Adam pretty much cuts himself off from Crow Creek. Miles West talks about the urban invaders with contempt as "people like them" who are "not desirable" compared with Crow Creek's farm or ex-farm families of the old days.

Pierce and West, unlike the Mitchells or even Mary Pierce, flatly refuse to adapt to change, an attitude likely to grow even more rigid as change intensifies and they grow older. Here I'd like to go back to my original theses: Culture is an inherited design for living. It has an economic basis, above all, how we get our food. It has universality, with family and property as its main institutions and religion at its core. Science, or its practical application as technology, is the main agent of change, since it changes the economy on which culture is based. In this book I'm going one step further: there is no substitute for the rural base of any urban culture, including our own. You've got to have so many farms and farming communities, or everything starts going haywire in the whole society. When Adam Pierce goes on about working mothers or Miles West goes on about rampant drugs and the erosion of honesty, I take some of what they say quite seriously.

In 1859, I mentioned early on, my great-grandfather and his family came out from New York State to help found, with other Quakers originally from New England, what they felt would be a purer, simpler society. Quaker founder George Fox ("O then, I heard a voice . . . and my heart did leap for joy") was a mid–seventeenth-century village cobbler and sheepherder who wanted to forsake the established Oxford- and Cambridge-educated urban order of English clergy and go back to the simple beliefs of the rural peasantry. My great-grandfather's fellow settlers named their small Iowa farming community after the Quaker poet John Greenleaf Whittier, famed as the voice of the preindustrial New England villager and farmer. Well into this century, Quaker historians were describing Whittier as preserving the eccentricities of mid–nineteenth-century Quakerism. I used to bicycle over to Whittier, just 3 miles west of Crow Creek, as a boy to see my mother's elderly Quaker cousins. I can still see the consternation of these goodly people when, coming back as a young man, on one visit I pulled out a cigarette.

In 1990 I went back again, thinking if any Midwestern farm community would keep to its old ways it would be Whittier. (Crow Creek, in Whittier eyes, has always been full of sinners and beyond the pale.) Sarah Pierce, Adam's mother, told me she expected I would find it full of old people. "Those Quakers seem to live longer," she said. "I remember so many that must be pretty well up in their 80s. They're quiet and peaceful and I think they live longer on that account. When I was young and growing up, the Quakers wore black clothes, you know. Black hats. They were enterprising too. So many raised strawberries. And one of the farmers sold ice cream, sold it down even in Stone City. Another peddled cheese. They'd do something like that. And Quakers who went out from Whittier did real well."

The farmer I went to see, Robert Williams, was sort of a distant relation.

His father, an orphan from New York City at the turn of the century, was adopted by a Quaker great-uncle of mine. The father became a Quaker too, took the family name, and passed it on to his son. Robert Williams's 200-acre farm a mile north of Whittier looked as old-fashioned as Alfred Zuckerman's did in Prairie. The barns were all painted and kept up, and there were cows, pigs, and chickens about. Williams, a lean, gray-headed man with narrow shoulders in his 60s, had a humble and mild expression. His wife, Maxine, a onetime nurse, looked to be a pleasant-faced, substantial, pious woman. They both looked pleased when I said their farm looked like something out of the 1940s.

"The school kids used to come out in the spring," Robert Williams said, "just to see what an old-fashioned farm looked like. We've got little pigs and chickens. Counting calves and cows, we've got about ninety head of cattle in all. We were in the 'Grade A' dairy business for years. Now the milk house is just sitting there. It was mechanized. We had milkers and a bulk tank where the milk was pumped out and taken to town."

The Williamses, in their late 60s, drove me up the road a short distance to show me the large white farmhouse my great-grandfather built in 1859. I was amazed it was still there and looked so well maintained. "One man had it for years and years," Robert Williams explained. "He died and the people who bought it wanted to fix it all up as a country place. And they put lots of money into it. So it's all fixed up inside with carpets and an open stairway. They wanted $60,000, $70,000 for it. Five to 7 acres with it." People who work in Cedar Rapids bought it. Williams said at least seven bedrooms were upstairs. I feel a pang of guilt for not trying to buy it myself. But what would you do with it? Heating a fifteen-room house through an Iowa winter would cost a fortune; the upkeep would be enormous. I'm just thankful people are still living in it.

When we got back to their farm, I noticed a big woodpile outside the house. "We heat the house with it," Williams explained, in his stolid, calm-mannered way. "That's just about the only past practice we see coming back. I don't know. I have always said that this farm would never support another family like it's supported us. The next generation's going to have to have a city job and then farm the ground on the side. You can't live off a farm this size. *We* have been able to, but the next generation won't. A lot of our young people commute to jobs in Cedar Rapids. Now our sons, one works at Cargill and the other at Square D. So many commute from small towns and the country."

He says he never takes part in any government program. Since Adam

Pierce finds he can't afford to stay out and even somebody as independent as Alfred Zuckerman takes subsidies, I ask Williams how.

"We're surviving. Nope, we don't take crop loans. I don't believe in government programs. I wish the free market would be left to work." What they do is to feed much of their corn and grain to livestock. "Oh, there's a lot of people who wouldn't be farming if it wasn't for government payments," Williams goes on. "That's the way they farm, period. I'm a realist and I can think of a lot of people that you take the farm program away and they could not continue farming. I'd gradually phase price supports out. I wish supply and demand would come into it again."

Is 200 acres enough?

"I've never been interested in renting land," Williams says. "We chore." I gather by that he means he and his wife prefer to make much of their cash income from livestock, more labor intensive than crops. "Our income has come from doing chores. With livestock. Rather than farming a lot of ground. This is our approach. We've had chances to rent ground many, many times over the years that we've farmed. But we always thought we wanted to take care of what we had at home. We weren't interested in going up and down the road, farming other ground. Concentrate on that. There's many ways of doing things. We'd just concentrate here. That was our way."

He speaks with great finality. "We use our old barn. Beef cattle. And hogs. We've got little pigs and now we've got calves. Then the barn's always clear full of hay. Of course this time of year it's getting used up. You could say this is a traditional farm. This is the way we know best to do. Some of these young folks, like that Clint Houseman, they're hard workers. He seems to be doing very well. Some of 'em farm real big. But some of them have fallen on hard days. You take Miles West. Expanded too fast."

The Williamses have known hard days too. Maxine Williams, trained as a nurse, says one winter she went back to work. "Times were really bad. I didn't even tell him I was going to do it. And I was gone too much. The children were too small. It was just a part-time job and we had baby-sitters. And we had pigs coming and he says, 'This is it. You can't go back to work. I need you here.' So I've been here ever since." Her words gave me a sense of thick, gloomy Quaker austerity. Whittier exemplifies a lot of the old tribal restraints, religious conventions, and patterns of obedience to authority you get in old traditional societies. Robert Williams, like Alfred Zuckerman, uses the collective *we* as if he and his wife have uniform views; I wonder if they really do.

"Well, we don't get up as early as we used to," Robert Williams says, as if reading my thoughts. "Six-thirty. Mother and I will chore an hour or so before breakfast. We feed all the livestock we got. Oats and hay and ground-feed. We buy protein. Feed them from calves up until they're ready to go to market. And feed 'em ground ear corn. The hogs get ground oats. We just buy the protein. We just have a little bit of everything. No sheep. No horses. We kept riding horses when the boys were children. Seems to be more horses around today than ever. But not for farming. If it wasn't raining today, I'd be out hauling manure. And we plow down legumes, sod. We don't use as much chemical fertilizer as some people do."

Is Whittier still a Quaker community, I ask. I noticed coming in that the meeting house is kept well painted and in good repair, and the graveyard beside it well tended. "It is and it isn't," Robert Williams says. "Would it be 50 percent here?" he asks his wife. "I don't know if there'd be 50 percent," he goes on. "The young people don't come to meeting. They go their separate ways. My own children have gone their separate ways. They will come to meeting once in a while. They have this background. There's ties there they don't want to let go of." It was a good way of putting it. There were ties, rural ties, a good many Americans instinctively didn't want to let go of.

"I think we lost our Quaker school in the 1940s," Williams says. "I have the records of when they sold the schoolhouse. It was moved away and made into a house on a farm. The older Friends in Whittier today, they still use *thee* and *thou*. When I was young, I was sent to a Friends boarding school in Barnesville, Ohio. We'd never used *thee* and *thou* at home and the school superintendent kept asking me, over and over, 'What did thee say, Robert? What did thee say?' At last I got the message; he wanted me to say *thee*." As I knew from my mother, the practice went back to the seventeenth century when the first Quakers scorned the newly fashionable *you*, used by the royalist cavaliers to show respect to single persons of rank; the Quakers clung instead to the *thee* and *thou* of their peasant villages, where an older English was used in rustic speech. And here it was still surviving among the old people of this Iowa community over 350 years later. It was a good example of the tenacity of some inherited culture.

"Our meeting is still Conservative," Williams says. "Oh, the men and women no longer sit separate. That went out about the time we were married, 1948. We think we're one of the first couples that came in and sat together. They thought it had to be divided at one time. For business meetings they even closed it all down, put wooden partitions across the middle, men on one side and women on the other. And then they had a slot in the door. There was an elder that'd sit there and they'd push business papers through this slot. And the men would work on it and they'd push it back to the women. That slot is still in the partition down at the meeting house." I ask what he means by *Conservative*.

"*Conservative* means it's a silent meeting until someone's inspired to speak by the Inner Light. They just get up and offer a prayer or make some quotation or talk about something in the Bible. It's interesting. There are still new people joining the meeting. There's a man who works at Rockwell; he and his wife come to meeting quite regularly. He travels. Every time you turn around, he's been to China or somewhere for Rockwell. But we don't have very many young people at meeting. It doesn't hold their interest."

He says the fundamentalists had attracted some of the Quaker young people. "I don't think in the future they'll be quite as strong. All this scandal. That Tammy Bakker, she's got false eyelashes that don't stop. A friend of ours, quite well-to-do, she was here in the community, isn't any more. Billy Graham got a lot of her money. Oral Roberts got some too."

Maxine Williams complains, just as Mary Pierce does, that too few of the newcomers moving into Whittier want to be active in the community. "Those people from someplace else, they really don't care whether they're involved or not. Most of us have always felt that if there's something going on in Whittier, we want to take part in it. Someone dies, everybody pitches in and helps that family. You know most everybody. Quakers are only normal. Maybe that's what happened to the young people. They became too worldly and so they're not Friends anymore." She turned to her husband.

"Commuters like their weekends," he says. "Our grandson, he's involved with the Little League in Crow Creek." Robert Williams shakes his head. "We don't know what a weekend is. Well, my sons are just not interested in farming. And there's no use making a fuss about it. Maybe one of the grandsons will be interested. Oh, I tease 'em. I tell people the biggest disability my sons have is their hands won't fit around a pitchfork. And chores, you know. They didn't care for chores. They didn't like to milk. Now if you ask a person in Whittier what they do, more likely than not they'll say, 'I work at Rockwell,' or 'I work at Cargill.' Places like that. They're very respected in the world."

Is there a price to pay for leaving the old ways? The Williamses, like most older Midwesterners, say there is. "Believe it or not," Robert Williams says, "Whittier has problems with drugs and alcohol. There was a bad period. And we're wondering how our grandson is going to handle it. One's just a sixth grader, but the other will be a freshman this year. And he's real easygoing. 'Whatever you guys want to do is fine with me,' he'll say. He's well liked and he likes people but he has a tendency not to say, 'No, I'm not going to do that.' He goes along with the crowd."

Whittier had a bad drunk-driving accident not long before my visit. "It put a different light on it for the kids," Robert says. "A very well-liked young fellow was killed. The boy driving the car was arraigned this week. Very remorseful. He got out on bail."

The best change for good in Whittier in the past 10 or 15 years, the Williamses agree, is the way so many young couples with children have moved in. But they do not join the community in the old way. "Society was different 30, 40 years ago," says Robert Williams, "because people did more things with each other in the small communities where they lived. Now because of cars and television, they don't. We used to get together in our community building once a month. Have speakers and socials. Now it just sits there. It's used some, but not much."

In a 1975 survey, "America's Third Century," in *The Economist* of London, its then deputy editor, Norman Macrae, one of the most acute observers of this country, wrote that increasing wealth and technology, that is, Robert Williams's city wages, cars, and TV, were allowing the inhabitants of rich countries to do something totally new in human society: namely, to live according to individual choice instead of in groups. I've always taken Macrae to mean small village-sized groups like Whittier, Crow Creek, or even Prairie.

Crow Creek's prime example of somebody who lives according to individual choice and hardly in a group at all is the much-praised young farmer, Clint Houseman. When I asked Clint what were his ties to Crow Creek, or those of his wife, Jenny, he replied, "None at all." Didn't he and Jenny ever get involved in anything? Clint grinned and said cheerfully, "Nothing. We don't go to church. We really have no ties to anybody here at all. Oh, I'll talk to them. If they're going by on the road, they'll stop and talk. Miles West and Adam Pierce go to church with some of my landlords. They know me mostly 'cause I rent some land from Adam and I farm and, you know, they probably talk among themselves. More than they'd really ever talk to me. Because we spend all of our time out here on the farms. I agree, the social fabric of a rural community like Crow Creek is really shot to hell these days. All our friends live somewhere else."

Clint, fair-haired, lean and fit and looking to be in his 30s; is just chaining up his dog, a ferocious-looking German shepherd, when I get to his farmstead. It is his twelfth year renting land around Crow Creek. In 1990 he dropped from 1,000 to just over 700 acres, deciding that was enough.

"One reason people talk is that we pretty much started on our own," Clint tells me over coffee as we sit around a big kitchen table. His handsome, sunburnt face, with its clear, clean-cut features and open expression, is distinguished by a broad grin, which from time to time lights up his whole face and makes him look much younger. Although both Clint and Jenny are in heavy work jackets and rubber boots, the living room seen through the doorway has more books and better pictures than you see in most farmhouses. There are copies of *The Wall Street Journal* lying about. Jenny, who soon goes off to wash up and run some errands, is even younger than Clint; she is pretty, with a rosy complexion, curly blond hair, and round blue eyes. I tell Clint it is very unusual today to find somebody like himself coming to farm from the outside.

"You don't have to grow up on a farm," he says, though most people argue you do. Clint says he had spent summers working on farms as a boy and in high school, so he wasn't entirely a city boy. "I didn't grow up on a farm. Not the whole time. In fact, I think it puts blinders on. It blinds you to opportunity. You know, if your dad raised cattle, you're gonna raise cattle. Or he raised corn and soybeans, that's what you're gonna do. And you take the farmer's kid and work his butt off from the day he can walk until the time he leaves the farm, you can burn him out. If you come from outside, you can figure out different ways to farm, ways more ecologically sound. No, I don't think you have to grow up in it."

As Clint speaks, his whole face expresses impulsiveness and enthusiasm one minute and a faintly worried uncertainty the next. He started out in 1978, he says, after going to Cornell College in Mount Vernon, less than 10 miles away from Crow Creek. He finished up at Iowa State University in Ames, where professors are urging him to come back and earn his Ph.D. in agronomy or plant science. He can't decide. He'll be 40 in a year or so, he says, although he looks much younger. "The drought set me back a couple of years," he tells me. "I can't afford to buy land yet. I started pretty much on my own. So to acquire the capital to buy land, there's just no way. That's one of the first rules I was taught—you don't buy iron and land at the same time. Your machinery and your land. So at this point in time I'm buying my equipment. My corn planter is new. And I bought a field cultivator this year. Equipment at farm sales went sky high. So I bought a new one as cheap as some guys were buying used ones at farm sales."

He says he and Jenny have been out working that morning with their herd of hogs. "Can you stand the smell of pigs?" he apologizes. "We were just out with them. We farrow-to-finish about a thousand head of hogs. Then we have a ewe herd. We have a calf herd. We quit feeding cattle last year. But we try to put most of our grain through livestock."

Does he use government programs?

"Oh, yeah, definitely. They still pay us, basically, about $2.40 a bushel for our corn. It used to be $2.90. That's one reason we have surplus corn. They're still paying us a price to produce it. Without it, well at harvest, corn was like $1.25. That's about half.

"We raise about seventy of our own cows. We increased the cow herd once we stopped feeding cattle. Cattle feeding's kind of died in Crow Creek. Soon we'll also have about two hundred ewes. We're cutting back on hogs. They're really high priced right now. Very profitable. We're culling back our sow herd because we get such a good price for the culls. Then we won't keep anything back until the price is cheap again. I can't afford to keep anything back. Right now we can get $55, $56 a hundred-weight for gilts. Sometimes they go down to $35, $36. Ken Bailey, over on the other side of Adam's farm, has about 4,000 hogs.

"I'm farming five different farms. Five families existed on this same land 15 years ago. The tax laws encourage large farming. We have to invest in the credit, we have depreciation, we have capital gains. You need to farm big just to write off certain types of losses. You can take outside income, turn it into your capital gains, and it gets taxed at a lot lower rate. Until the tax reform of 1988, somebody with a very profitable business somewhere else could put it into farming, deliberately farm at a loss, and even out his income. That encourages big farms. In your livestock, your grain, your confinement buildings, everything.

"Before, when we had inflation, if you owned land and pyramided the land, that is, you started cheap and added onto it, they really encouraged you to keep growing. So that in the late 1980s, when you had a deflationary period and then 2 years of drought, the people who built on their stair-step basis, somebody like Miles West, it's collapsed on them. It's now stabilized because so many big guys got their fingers burned."

Clint says farming is also hard, dirty work. "How many people want to smell like we do and work in the soil and get dirty? A lot of Americans want an easier way of life. Farming can be pretty clean if you don't have animals. But not many people even want to work as hard as a grain farmer does. As long as our society demands cheap food and the average family spends just 13 cents out of every $1 for food, most people are going to want a system that keeps it that way. That means the use of petroleum, herbicides, and big equipment. Maybe we're getting cheap food at the expense of our environment. It may catch up with us later. As far as our water being polluted. Our land washing away. Cancer-causing chemicals. I band my herbicides and a lot of my nitrogen. Or use postemergent sprays. If you band, you just put it over half the ground. You can cut your usage rate by 50 percent and get the same yield. By banding, it doesn't get tied up in the soil as fast. It's available to the plant longer. It cuts my fertilizer bill in half. Phosphates. Potassium. That's one thing your really big farmers can't do. They have time limitations. If you get too large, you start getting inefficient."

That day a blacksmith from Anamosa, a friend of Clint's, is out in the yard putting together a sprayer. "This field along the blacktop, I'm using a postemergence herbicide. I'm building a sprayer so it sprays right over the row. Which no one does around here. They pretty much broadcast and then forget about it for a year."

Clint says with livestock you have to go with the price swings. Right now hogs are very profitable. Eighteen months earlier, you lost money with them. He figures, very roughly speaking, hogs have a 3-year cycle and cattle a 10-year cycle. There were record-high cattle prices in 1988. "You have to understand the price swings. A lot of people, you know, they'll start them when it's bad. Or they start when it's good and then it goes bad on them. Fifteen, 20 years ago every farmer in Crow Creek had livestock."

I ask about Miles West's farm. "I don't know much about his operation," Clint says. "He owns some of his land. His dad farmed. It's more of a generational-type thing. He's got at least one hired man. Jenny and my dad help me here." Jenny, a Springville girl, though they'd met while skiing in Aspen, works right beside Clint in the fields, running heavy equipment just as Bud Petrescu's wife does in Prairie. Indeed, Clint and Bud Petrescu, like their wives, have much in common. They are the same age, both went to college, both love farming and work unusually hard at it, both rent large amounts of land and can't afford to buy any. Like the traditional farmer who resists change, if Clint Houseman and Bud Petrescu are any example, the younger, entrepreneurial farmer who welcomes change and tries to harness it is a fairly common type too.

Clint says the homesteads are abandoned on two of the farms he rents. One sees empty farmhouses scattered around the Iowa countryside, but far fewer than in North Dakota. In Iowa, where virtually the whole state is within commuting distance of a big town or city, a great many farmhouses are occupied by nonfarmers, something you rarely see in North Dakota. One farm he operates, Clint says, was the home place of Harold Fletcher. Until he died in 1988, Fletcher drove out from Springville every day, taking pride in keeping up the house and barns. He was an exception. "In Iowa," Clint explains, "the nicer the buildings are, the more property taxes you pay. So it's a real advantage to bulldoze your buildings down or let them look like hell. The barns get derelict. The worse they look, like that old wreck at Adam's, the lower you get assessed. The banks are getting so they just appraise the machine sheds and grain storage. They don't count the old barns as having real value anymore."

I observe there seems to be more and more grain farming. He agrees but says, "Look at those raising livestock. The number of farms with five thousand to ten thousand head of hogs on just one farm has really grown over the past 5 years. Someone of my size may not exist in a few more years. You can speculate that pork may become like chicken, thoroughly integrated. You could have five, six large companies pretty much controlling the inputs and all: feed, the genetics, marketing. That's what the chicken industry is right now. Look at your Tysons and your Purdue out of New York. Tysons is into Iowa; Monfort is into Iowa when it comes to buying hogs right now.

"I'm not saying hogs will necessarily go the same way. Hogs are harder to raise than chickens. Chickens, you just stuff a bunch of them into a cage and debeak 'em and you can pretty much forget about them. Hogs, you have a lot more disease problems, personality problems. Crowd too many together, and you get a lot of tail biting, ear biting. There's a lot more care needed. Plus, if the animal welfare people push us far enough, as far as what we can do with our livestock, that'll discourage corporate farming anyway. I'd rather they got after them. We use a lot of straw yet and pens. We don't tie a sow in a stall her whole life. We put our sheep out in grass. I keep sheep and cattle to graze our set-aside. Try to tie everything together."

One cold afternoon, after it had been raining all day, Clint asked me if I'd like to come out to one of his farms with him. He had to check on his cows. Luckily he had a pair of rubber boots and rain gear that fit, and so we spent a couple of hours trudging around the sodden, rain-swept hills. It keep drizzling the whole time, and as it got near dusk, the dim, chill wetness gave the Iowa landscape a wild and unfamiliar look, like the Yorkshire moors. It was great to be outdoors. I asked Clint how he happened to come to Crow Creek.

"I had a friend farming here," he said. "I came in with him. Then he became a Jehovah's Witness and lost all interest in farming. That's how I ended up by myself here. You know, I tend to stick at home, working. I don't get much into the community or hear the gossip or know what other people are doing." Someday he hopes to own a farm. "The farm crisis of the 1980s set me back 5, 6 years. Then the drought of 1988 and 1989 knocked me back a couple years more. I'd have thought I'd be doing a little better by now. Land prices got low, but to take the risk and buy then would violate what I call my Golden Rule. That's where a lot of people would get into trouble. They buy a lot of new machinery at the same time they buy a lot of land. And they don't have the income to keep up the payments."

Clint feels America's farm crisis, if history repeats itself, won't be over for another 4 or 5 years. "I really don't have a crystal ball. The average person may get poorer. With our balance of payments, we're living beyond our means. We either have to increase our productivity or lower our standard of living. I reckon we're living on borrowed time."

WE'VE SEEN HOW much Crow Creek's people vary, the old timers like the Mitchells and the Biggses, and farmers like Adam Pierce, Miles West, and Clint Houseman. Lastly we'll turn to the urban invaders, the Alvarezes, the Novaks, and even the Methodist preacher, Rev. Bachman, who happens to be from San Diego. As I said at the outset, most of the time they all go their own way. But now and then they have to do something as a community, as a group.

In Prairie, the county fair, like high school basketball games, brings everybody together. In Crow Creek, except for possibly Little League, it is the streetlights.

There are nine. The annual electric bill, going up every year, is just under $1,000. Since Crow Creek has no local government or authority to tax, how does it pay it? In 1990 the Civic League sent word to all forty-one households in town, plus a few of the nearest farmers, to come to a meeting in the basement of the Methodist church to decide what to do. In the old days the women of the Civic League would just have raised the money themselves. After all, it had been founded for that purpose in 1922. At first called the Improvement Club, the aim of its women members was to raise enough money to put a wooden sidewalk down on one side of what is now Pleasant Street, so everybody wouldn't get their feet so muddy when it rained. Their husbands provided the labor.

Next came streetlights. In 1925, once rural electrification got as far as Crow Creek, it was decided to put one electric light in each corner of town to replace the existing gas lamps. The operating cost was $4 a month, $48 a year. In 1990, 65 years later, it was more than twenty times more expensive. And these days, the Civic League itself has fallen on hard times. It still sponsors Crow Creek's annual Christmas tree, where everybody gathers to sing carols and somebody dressed as Santa Claus hands out candy to the kids. But the old family get-togethers—picnics, potluck suppers, ice cream socials—haven't been held for years. Just about all the Civic League does these days, now that the Little League raises its own money to buy equipment, is to sponsor a couple of visits a year to the Linn County Care Center, what used to be known as the county poor farm, a home for indigent old or retarded people.

The night of the meeting, people from fifteen households showed up at the Methodist church. Metal folding chairs were set out in the basement. Facing a crowd of twenty-five or thirty people, seated at a table in front of them, were the stalwarts of the Civic League—Alice Mitchell, Louise Fisher, Mary Pierce, and the current Civic League president, Cheryl Resnik, a pretty, fair-haired young woman with untidy curls on her forehead; her husband, Joe, sold cars in Cedar Rapids. In the audience were Adam Pierce, the Wests and several other farm couples who lived in or near town, the Biggses, and Mrs. Fletcher. Scattered among them were urban commuters, what Miles West calls "the new breed." Absent, though they too lived on the edge of town, were Clint and Jenny Houseman, but then they never came to anything.

Most conspicuous among the newcomers was a swarthy, short, thickset, Hispanic-looking man, who stood out strikingly among the farmers, with their weathered sunburnt faces and white foreheads once they took off their seed caps. This was, I was told, Pedro Alvarez ("Montezuma," a few of them called him behind his back), a Mexican-American who had moved into Crow Creek with his wife, Terri, a year or so earlier. They both worked in the office of Rockwell-Collins's Anamosa branch. Alvarez looked to be a man in his 50s, with a thick broad nose, a seamed darkish face, and bushy black brows. Long curly gray hair, in ringlets like a rock star's, hung to his shoulders; he wore a ring in one ear. When he crossed his legs, his bare feet in sandals also stood out among all the blue jeans stuffed into work boots. Terri Alvarez had a pale face, narrow shoulders, an anxious, slightly defiant look, and blond hair tied in a ponytail; she looked much younger than he did. I was told her son by a previous marriage, now 17 or 18, lived with them. Alvarez, too, was said to have left behind a brood of children with a former Hispanic wife in the city.

Another new face, a huskily built man of about 50 who was in a wheelchair, I guessed to be Tony Novak. In his movements, confined as he was, one detected the old athlete. He was with a woman with a thickish figure, whom I learned was Maggie Novak. She was about his age and faintly stout, her hair was straw colored, and when she spoke it was in a penetrating, hoarse, heavy-smoker's voice. But there was an immediate perceptible vitality about Maggie.

After Cheryl Resnik opened the meeting, Mary Pierce took over. She wanted to thank, she said very warmly, Mr. Alvarez for organizing the people on his street, which runs past the church. He got them all to chip in to help pay to put oil on the gravel and cut down on the dust. Moreover, entirely on his own, Mr. Alvarez took his pickup down to the Wapsipinicon River, loaded it up with gravel, and went around Crow Creek's streets, filling up the worst potholes, something nobody had done for years. Mary called for a round of applause for Mr. Alvarez. This done, Alice Mitchell stood up to talk about the streetlights. "Since Crow Creek is not incorporated," she said, "we have no mayor and no power to tax our residents. When something is needed, the Civic League has always raised the money."

A long discussion followed, with Miles West, like the leader of a government in power, and Maggie Novak, a scrappy opposition, having the most to say. At last it was agreed that Mrs. Mitchell would make up a form, get it copied, and supply envelopes. The form would ask each of the forty-one families living in the town proper, plus a few nearby farmers like the Pierces, who lived close enough in to be affected, to voluntarily pay $25 a year. It would come to over $1,000, enough to cover the cost. Pedro Alvarez, speaking in a gruff, husky tenor with a pronounced Spanish accent, volunteered to pay for the postage. No, he added in the tone of somebody who is afraid he won't be believed, instead of mailing the forms, he would personally deliver each and every one to all forty-one households. He smiled a broad smile; as a newcomer, he said, it would help him get better acquainted.

It took him three nights, setting out every evening just as soon as he got home from work. Since it was the first time anybody went to all of Crow Creek's households, it was kind of a poll. As soon as Alvarez finished, I went to see him to ask how it had gone.

"Some of 'em were belligerent!" he erupted, even before I had time to sit down. We went out to the kitchen where Terri Alvarez was making coffee. "Some of them felt they don't *belong* to this community," Alvarez went on.

"What was the worst reaction?"

"The worst? Oh, they'd complain there was other projects, like our garage sale or our supper sale at the school, and that money should go for the lights. Oh, everybody talked to me. I'm an easy guy to talk to. Everybody welcomed me. When I say they were belligerent, I don't want to get anyone in trouble or anything. But they rather raised their voice, yelling like 'What does the money go for?' 'Where'd you spend that money from the supper? I thought *that* was for the streetlights.' They didn't even *go* to the supper. So why are they criticizing if they didn't take part in the first place? If tickets were $2 apiece, we coulda used the extra $4. That was for the school anyway. I believe if you're gonna complain about something, you better know what the hell you're complaining about. Seven cents a day is all the lights amount to. Approximately.

"One guy flatly refused to give. He says, 'Well, you can give me a form. But we're not part of this community.' I mean their house is right on the edge of town. It's a technicality. Anybody can get mail through the Springville post office, not Crow Creek. All you do is tell the post office and put up a postbox. But I believe, if you live in this town, patronize it. We want our post office to stay open. If they have kids, you're not going to tell me that that child, when it gets 10 years old, it's not gonna walk a couple of blocks to come in and play with its friends here."

"Just one refused?" I asked.

"Just the one. They just flat-out-assed refused. The other, I don't know, he raised his voice. Both of 'em raised their voices. The man and the woman. And I says, 'Wait a minute, now, wait a minute. Let's get somethin' straight here. You don't *have* to give for the streetlights if you don't wanna give. Just sign the paper 'No,' and send it in. That's all you have to do.' I had a form and then they're supposed to return it to Alice. But I don't think they had any right to take their vengeance out on me. They shoulda just wrote down on the paper 'No,' if that's what they wanted. I got a little irritated myself."

I asked him why he didn't just mail them.

"I was gonna. I volunteered to pay for postage. But, I says, I'll take 'em around. It was kinda funny too, because the first night I started deliverin' these forms, I made it to *one* house. I like to talk. I think in a small community like this, people are more willing to talk than they are in a city. Sure, you have your next-door neighbors there and you can talk. But you don't have that closeness that you get in a small community, you know."

Alvarez said he felt everybody should be happy to chip in for street-lights. "Even if you don't have children, you should still donate. With a lighted community, you don't have the vandalism and theft and bodily injury. What if it's pitch dark and somebody goony drives down the street? Now some of the lights are county-paid. Any light on E-34, that's the pavement going west. I think there's three lights. That bad turn got one too. But if these lights were the only lights on, the rest of the town would be black. And kids play out there. You can hear kids playing out there ten, eleven, twelve at night. You don't have to worry so much about them. If they have lights. You can kind of watch them, you know. I watch my neighbors' kids, keep an eye on them. I like the lights. I'm all for 'em. Two dollars and seven cents a month. Twenty-five dollars a year."

In 1988 the Alvarezes built a big swimming pool; Crow Creek, which had none when I was a boy, now has four. In the summer kids swarm there, just as we used to go to the creek. Alvarez finds Crow Creek very neighborly. "All you have to do is say, 'Hey, would you mind helping me a minute?' They flock like flies. All you gotta do is provide a beer. Like help on a car, or you're building a building, or painting something. There's definitely a strong sense of community here. In the Little League, the kids come from way out in the country. You oughta see Crow Creek on Halloween. We get triple the kids that we got in town. The farm kids come in because their parents know there won't be any vandalism or any damage to the candy we give them. We got good people here.

"You know that couple that was outright belligerent, when I was pass-ing out the flier? It was like they felt they don't belong. Like people don't talk to them as much as they should. They don't have children. And they criticize other people's children for doing things that don't mean nothin'. I have children coming into this yard here that do minor damage. But I don't say nothin' to them. I just fix what they damage. And I think they feel—this belligerent couple—they feel they don't belong in Crow Creek or want its kids anywhere near 'em. They don't like children in their yard. Just cutting through their yard. Sliding down their hill. Something like that. I think they feel they don't belong, and the reason they don't belong is their attitude. You're not going to belong to any group if you've got a piss-poor attitude and don't take part. The way to get people to accept you is to be friendly, open, and honest."

He said Alice would know who gave and who didn't. He himself didn't want to know. He explained, "You see, if this couple, this belligerent couple, doesn't give and that's brought out somehow, it'll make it worse for 'em yet.

"The other reaction I got taking around those streetlight fliers was outright bullshit. They just loved to bullshit. You know, talk, talk, talk, talk, talk, talk. I had to fight to get away. They were really, really friendly. And people just love to talk. So I ran into three kinds of people in Crow Creek on the streetlights: Flat-out-ass no. Belligerent. Or out and out friendly."

I asked about the Pierces and the Wests, but Alvarez said he hardly knew the farmers who lived outside the town. "They drive by and we wave. From what I hear, I'm thankful for 'em. They keep this town, in a sense, going. They take part, keep the church going. I'm glad it's there. I'm Catholic. We don't go. It's a good place for get-togethers like on the streetlights. I like a church in a community."

He had his own idea on what would bring Crow Creek's people together. "Okay, I would say, my idea for Crow Creek, probably not the best, and I'm thinking what they ought to do is have the real heart of the Civic League, you know, the old guard, meet every so often, once a month, something like that. But when they want the community to meet and voice opinions and participate, they ought to have a big . . . beer party! You could have coffee and pop for the old-fashioned ones. Maybe I shouldn't of said that. Maybe I didn't think too much. When you say *party* to me, I go for the six-pack. I just think people have to visit. It's what holds them together."

Terri Alvarez didn't think much of the idea. "Would you have Jonas and Alice coming to a beer party?" she asked her husband. "It wouldn't work for everybody. You can't have a beer party for Crow Creek. Sure, we like to drink and a lot of our friends here in town like to drink, but not everybody. Kenny could come but not Louise. Joe might come but Catherine wouldn't. There's very few older ones that come to any of the beer parties around. You'd get the Evanses and Jimmy and Anita and Cheryl and Bud and Lester and the Clarks over there and Sue and Jack. . . . When we have parties and stuff, it's not that the older ones aren't welcome. They can come. But it's the younger ones who do. I don't know how many times we've sat out here on the deck in the evening and by the time it ends up there's six or seven housefuls with us. It runs in streaks. They see us sitting out and they'll come over. Pretty soon somebody else comes over. All you gotta do is provide beer and you have a party."

AS STRONGLY AS its natives resist changes, Crow Creek has grown a lot more open to outsiders over the years. Maggie Novak still recalls a brick shattering the front window when she and Tony first moved in nearly 30 years ago. "At night. Because we were Czech, Democrat, and Catholic, 'Boheemies,' as they say. It was like the old South. I'm not kidding ya. It was terrible. We called the sheriff. He said, 'There's nothing I can do about it unless they break into your house. All I can advise you is to get the heck out of there.'

"I told him, 'I ain't never going to move out of here.' Sure it was kids. We *hope*. But they *heard* this from their folks. Children hear this sort of thing at home. Oh, old Barney Fisher came over and just raised all kinds of hell with us. Because we were Democrat and Catholic. He just walked in. Didn't knock on the door or nothing. He started in on my politics. I said, 'No one asked you here. There's the front door. Get the hell out.' He stood there for a minute and I took hold of his arm and said, 'I'll help you!' " Maggie gives a hoot of raucous laughter and lights another cigarette. She is seldom without one, taking a puff for a dramatic pause, or stabbing the air with one if she wants to make a point. "I was born and raised in Anamosa," she declares, as if that explains her edgy style.

The streetlights issue, I should interject, was resolved when all but a few families agreed to pay $25 a month. It is good to have Crow Creek's old general store, closed for years, back in business as I mentioned; even the gas pump in front is in operation. Rod and Angie Johnson, a once-local couple who went to Florida and came back again, run it almost like a neighborhood club. A pot of coffee is always going, and they've put in a table and some chairs and sell homemade cinnamon rolls and sandwiches. Maggie tells me Angie Johnson put a jar out on the counter, where people can donate change to a streetlight fund; it aims to put those who refused to pay to shame. Or so Maggie says. She talks with lively animation, her laughter, her assertions, her gestures gaining in gusto as she goes along. Alvarez might want to keep the delinquents secret, but Maggie knows them all, raising her eyebrows in despair at their cheapness as she reveals all their names.

She leans forward conspiratorially. "Did you get that at the meeting in the church when Jonas said, 'Okay, say there's people in Crow Creek who won't pay $25 a house. Would you be willing to pay $60, $70?' Jonas asked that. He and Alice said no. Mary said no. *We* would have been more than happy to pay. Tony had already talked to Ed Biggs across the road. We have one light right out there. They agreed, if need be, they'd take that light over and each pay half.

"It's like the 'Neighborhood Watch' signs in town. *We* bought our own. Now I've turned in four or five petitions trying to get the gravel road out there oiled so we don't have so much dust. They won't do it. Alvarez got his oiled. They went together and did it in front of the Methodist church too. We tried it in our street but so many people won't do it. Ed won't do it because he says it costs too much. Same thing with Louise Fisher. The old people have always lived with the dust. I think they've just sort of grown used to it. And they don't have the kind of ready cash the Alvarezes do, with both of them working at Rockwell-Collins."

Maggie raises an eyebrow in disdain and lowers her voice. The Alvarez house, she says, built before they were married, was subsidized by the government. Terri, as the sole supporter of an underage son, was eligible for a low-income housing loan. "I don't know how it works since they got married. That's how it was. I can't believe it myself." She makes an incredulous face. "When Mary said that night how nice it was for Pedro to pour that little load of gravel on the potholes, I thought, 'Hell, I'd gravel the whole town if the government built a new house for me.' I didn't say it but I thought it. To me it's a rip-off!"

Maggie gossips with great enthusiasm, eyes flashing, eyebrows raised, cries of indignation, bursts of withering sarcasm, the conspiratorial undertone, the grim judgment, the dramatic pause to puff on a cigarette. "I like Pedro and Terri," she cries. "But I couldn't live like that. I suppose you've got to take the bad with the good. Compromise. Pedro could do a lot for the community. Going around and collecting for the streetlights. Stuff like that. Except when they're having beer parties on their back deck. Tony and I were invited over to a party." Her voice falls. "And half of 'em were smoking pot. Young people in their 30s. *We* came home. All I'm saying is, these people all party together. Frequently. As long as it's just beer, I say fine. It's not destructive as long as they don't do it every night and raise a lot of hell and keep the neighbors awake. But Pedro's wrong when he says a beer party brings everybody together. It just brings their *group* together. All the people in Crow Creek the same age who do the same things."

She shakes her head back and forth. "Hardly anybody goes to church. Maybe none of them. Maybe they sleep on Sunday. From the night before. Sometimes it goes on until three o'clock in the morning. That time we got invited, that went on for 3 days. It lasted for 3 days. Just playing volleyball and swimming and drinking and some of 'em smoking pot. Having a good time. It would quiet down from about two in the morning until seven or eight. Then it started up all over again."

What struck her most about the beer parties, Maggie says, was that those who went saw nothing wrong in them. "They think it's respectable because it is *in their group*. You heard Pedro say that what Crow Creek really needs is a big beer party? That's his answer to everything. Have a great big party and we'll get together and all get drunk. They kinda look at, well, Tony and I even as kinda stodgy. Oh, they definitely have good intentions and they're likable people, but you've got to ask: what's the effect on the kids? Most of their children are little; who knows? Two of the boys, teenagers from two of the families, they've been in all kinds of trouble all the time with the law. One's been thrown out of school. He was on probation. The other was caught twice a few years ago breaking into the Crow Creek school. Vandalized it, wrecked a computer or two, wrote on the walls. He was just 14 or 15 then. What they're finding out, if you read any articles at all on it, the reason these children are having the drug problems they're having today is that their parents are doing this in their homes. The parents are openly using drugs and all this stuff. And telling their children, 'Now don't do this. You're not to be a pothead.' But here sits the parent.

"Then they set the kids in front of that stupid television. I think the effect of TV on children is terribly underrated. But children really learn the basics from their parents. If the parents say, 'Don't drink, don't smoke,' and they do it themselves, what then? Children learn from examples their parents set. If parents, out in the world, look honest and forthright, but behind their backs cheat, lie, and steal, the child will think: well, to cheat, lie, and steal is all right as long as you keep it hidden. It's still parents who teach their children." The Novaks raised four of their own.

As she talks, Maggie has a way of looking you straight in the eye. Full of life, opinionated, harsh, and loud, she is just as stridently critical of Crow Creek's more traditional element, the older people downtown or some of the farm families just outside. "They're always looking over their shoulder at the Ten Commandments. Tony had a grandmother that way. Irish. Strict Catholic. Nobody could talk about divorce. Nobody could take a drink. It was really something. Just before she died, she really loosened up. You know, a couple of her daughters had gotten divorced. She had to face it that times had changed. It used to be she'd really preach, 'This is wrong! Don't do this! You'll go to hell!' It's still like that in Crow Creek."

One time, she says, wetting her lips, she suggested the Civic League members charter a bus into Chicago and go to its annual flower show. "They said, 'Are you out of your mind?' They were scared to death to go into Chicago. What is it, 4, 5 hours? So much of their life is lived in this one little town. It's like, you know, they're afraid to venture. To mix with anything else. Some of those women, they just flatly said no, it was absolutely out of the question. Well, one year, I guess, they *did* finally charter a bus. Alice finally got them to do it. And they went into the flower show. I think being a woman in a small community and everybody knowing everybody else and all this stuff is fantastic. But you've got to live beyond your community. There's a world out there beyond Crow Creek." I knew what she meant. For some years in old age my mother lived upstairs in the Adam and Mary Pierce house. When I used to visit her in the summer, we had to be ready to hide our beer glasses if one of the older people came to visit. The Crow Creek store didn't sell beer until well into the 1970s. We used to have to walk down the railroad tracks to the store in Stone City, a mostly Catholic community.

Aside from people having coffee and making conversation at the store, just about the only place where many of Crow Creek's people come together is on the baseball diamond. I ask Tony Novak, the main organizer of the Little League, about it.

"I think the reason it's successful," he says, "is that we don't have any power struggle there. We're a democratic thing. The older people on the farms, the ones with the power push, they're not into it much, maybe some of the grandchildren. And the young kids that just moved in, that's smoking pot, living together, they're not into it either. Almost none of their children take part. Like the couple next door. They aren't married. A young man and a young woman. Both have jobs. They live together as a family. No children. And it's the same, one house across from the schoolhouse. Man and woman live together. No children. He has a girl from a previous marriage, and she has a boy from a previous marriage. They've got the boy with them. They live together, just set up like man and wife."

He says all the other kids in town were active. "The parents may come and watch the game. The mothers do a lot. But the fathers don't, except those that coach. Right now the field up there has become the social spot of the town. If any of the kids gets bored, they run up to the diamond. So there's always a game to watch or people to talk to up there. People come up just to see who's up there. Anybody who has a family reunion in town, they end up there, on the diamond, playing softball. Our family is always home on weekends. And they all go up and play ball. It's getting better every year. Because the children are involved, the parents come. Last year we had a meeting up here at Crow Creek school, trying to get parents to become involved. I think we had 25 people. This year we met up at the gym and had 125."

Tony Novak, a former athlete himself, had both legs amputated after being hit by severe diabetes. He uses artificial legs and goes around in a wheelchair, but with strong shoulders and arms he keeps remarkably active. Since he gets disability payments and is home all day, the Novaks' yard and swimming pool is a popular place with Crow Creek's children. "The kids and I built the bleachers," he says. "Out here in the front yard. Then we hooked a tractor onto 'em and drug them up there to the diamond. That little stand was an old shed somebody was gonna tear down, and we drug it up there too. I talked to the electric company and they've put up a light pole." The stand, which has an electric icebox, sells cold drinks, candy bars, popcorn, bubble gum, potato chips, and hot dogs. Maggie found a used microwave oven for $35. The stand made $2,800 in the summer of 1989 and over $3,000 in 1990. The Novaks say there were forty-three games between the end of May and mid-July in 1990. The money goes to buy equipment and uniforms. A good bat costs $25 and lasts 3 years.

"Right now," says Tony, "we've got Minor League, which used to be Peewee, that's kids 6 to 9. Then there's Little League, 9 to 12; we have two of those teams, up to twenty-eight kids some years. Now the 9-year-old, if he isn't mature enough for Little League, he can stay back another year and get more experience. We also have girls' softball, grades third to sixth. The older boys can go into Babe Ruth, which is 13-, 14-, and 15-year-olds. Then we're done with them. But a kid can start out here when he's 6 years old and keep playing with us until he's 15. The boys play baseball, the girls softball. We've got five coaches and five assistant coaches." He says the children come from Crow Creek, Anamosa, and the two small neighboring communities of Stone City and Martelle.

One bad thing for kids today in Crow Creek, Novak says, is that the old railroad tracks were pulled up and the land sold. The man who bought it fenced it off. "It's all gone," Tony tells me, "the whole railroad bed. They ripped out all the bridges, the rails, the ties, everything. All the way down. It's all growing up in wilderness." It is a real desecration. I spent so many summers as a boy, like everybody who grew up in Crow Creek, going down the railroad track, exploring and fishing and swimming. Everybody kept herds of milk cows in those days, so the pastures on both sides, with the creek meandering back and forth between them, were like parks. You could walk under the limestone bluffs all the way to Stone City and the Wapsipinicon, even up to Anamosa.

"All the kids from Crow Creek, my kids growing up, they always went down to the railroad track." Tony shakes his head. "My boys camped down there and swam down there. They pret' near lived down there all summer. It makes you wonder." He says he still finds Crow Creek an ideal place to live. "That's the reason we came here in the first place, just to raise our kids in the country and in quiet surroundings."

Maggie Novak isn't quite as enthusiastic about Crow Creek. "If you don't happen to be Methodist," she told me accusingly, "no matter how long you live here, you don't become part of the community."

IS IT TRUE? Anthropologists tell us religion, the world over, is the core of any culture. Civic League, the old PTA at the school, like the Little League now, and even the Alvarezes' beer parties, or a coffee break down at the store, bring people together. But like people everywhere, many in Crow Creek feel there has to be more to life than just living from day to day. They look for a deeper meaning.

There is no doubt, I think, that farming leads to an agricultural moral code. "Do unto others" almost certainly has its origins in the mutual-help ethics of early agriculture. Such values as paternal authority, early marriage, divorceless monogamy, and multichild marriages, all much stronger in rural communities like Prairie and Crow Creek today than in cities, and within Crow Creek, as we have seen, stronger among the farm families than the city commuters, all grew out of the imperatives of early farming. So do most of the great religions; prophets come from villages.

As we move away from farming, with its annual rebirth of life and the mystery of growth, our beliefs naturally tend to weaken. In the modern city we see that men can walk on the moon or blow the planet up. We are beset by so many conflicting claims and ideologies, most of them appealing to scientific logic. We learn the Bible is more myth than history, to say nothing of the divine revelation our grandparents thought it was. The way to success in the future, we hear, is not in growing or manufacturing things so much as in mastering semiconductors, supercomputers, robots, biotechnology. In Crow Creek, less so in Prairie, holy days become holidays; the highways are crowded, the churches half empty. Where once it was David and Goliath, and Moses in the bulrushes, now it is Batman and Teenage Mutant Ninja Turtles. Half the appeal of rock music, with its heavy drumbeats and screaming guitars, is the way it drowns out thought with its ear-splitting, all-obliterating din.

As it happens, Crow Creek for a time had a musician, a Californian, as its minister. Rev. David Bachman, tall, broad shouldered, and stout, with a beard as black as ink, a jovial man still in his 30s, saw his rural congregation through urban eyes. Crow Creek's Methodists, I think, like rural people everywhere, take their religion pretty much for granted. The promise of salvation, the ideals of forgiveness, charity, and love of one's neighbor, though preached about in the Sunday sermon, are not things they submit to scrutiny in their everyday lives. Rev. Bachman, in his own way, subjected their values to a great deal of scrutiny. Had Crow Creek's churchgoers known what their preacher really thought—the few sermons I heard him give at the Methodist church were nothing like the views he expressed privately to me—I think they would have been rather shocked by his original and radical ideas. It never came to a head, as Rev. Bachman didn't stay long but was moved, as has been Methodist custom since its circuit-preacher days, to another parish before he ever made these ideas known.

"It's interesting," he tells me in his deep bass voice when we meet one day for lunch in a restaurant in Mount Vernon, some miles from Crow Creek, "to read essays written in the 1740s and 1750s when the Industrial Revolution began. How it was going to destroy the religion in the world. Everybody said the country people will lose their faith if you take them off the farm. It's not logical. You know traditions are only a way of dealing with your environment. If the environment changes, you are left with traditions that aren't useful anymore. This is happening now. Everybody's afraid Crow Creek is dying. And they want to keep it going."

At this time I had not yet interviewed the bishops of Durham and Southwark in England who told me, as quoted at the outset, that urbanization was directly to blame for the present decline of the Anglican Church. This does not happen right away. As Ronald Bowlby, the Bishop of Southwark, said, the people who flooded into the English cities in the nineteenth century from villages brought with them an ingrained religiosity, and it is this, at the end of the twentieth century 100 years later, that "has finally drained away." Those essays written in the mid-eighteenth century were right; the erosion of rural-instilled belief just took a very long time.

To Rev. Bachman, the fears of Crow Creek's old farm families are justified. "There are all the overt signs the old group as they knew it is dying. You've got children leaving. You've got new people coming in who don't understand the rules. You've got farming as a profession changing. There's two ways to go: you can either adapt—which practically never happens—or you can build rigid defenses against change. A classic way is to retrench, to make the old rules even more rigid. And of course the sad thing, the pathological thing, is it makes it worse. Until it explodes."

Isn't he, I ask, looking at it from an urban outsider's point of view? "Okay, I'm a depraved Californian," Bachman says. He had been a professional musician, a pianist, he says; he taught music at the University of San Diego and he played church music around town. He'd come into the ministry fairly late in life. What fascinates him about Crow Creek, he says, is how "dysfunctional" it is getting to be, like some bodily system whose functions are going haywire.

"In Crow Creek, three families dominate the political life of the Methodist church: the Wests, the Pierces, and the Scots, and Jane Scot is Adam Pierce's sister. The Pierces sing on Good Friday and the Scots sing on Easter. It's just amazing the elaborate scenarios that are played. And they work. Everybody knows they work and they work. Well, they get some sort of emotional payoff, psychological payoff. In Martelle, my other church, the congregation is not as inbred. It needed to go with somebody, hook up with a part-time pastor. Crow Creek couldn't support a pastor either. Put them together and they do fairly well."

Bachman says he has been in Crow Creek and Martelle, where he and his wife have lived in the parsonage, for a year. He is still finding Crow Creek enigmatic. "On the one hand, it seems quite backwoodsy, with all the images that brings. On the other, there's all this evidence to suggest the contrary. Clint Houseman hedges hog prices on the Chicago Board of Trade. Some of them have been all over the world. These are intelligent people. Old Mrs. Pierce, well, she's in her mid-80s now, but she's a very intelligent woman who never had a chance to use her brains. She sat there and read, read all her life. And watched the little rural world around her gradually go to pieces." I tell him Sarah Pierce has letters from her New England Quaker ancestors going back to before the Revolutionary War. They mean a lot to her.

"It's a problem ministering to congregations like this because they confuse the spirit of something with its physical essence," he says. "If that makes any sense. It's the animistic side of society. And I've never been anywhere where power-broking was more blatantly obvious. The old ones aren't involved too much anymore. It's the children who have inherited some of these agendas. Miles West is a power. He's used to giving orders. He's able to be the power broker in the church because nobody wants to risk his wrath. The administrative council meets once a year and it's ruled by an iron hand—Miles's.

"There's fifty a Sunday there. That's what they say. I don't believe it. In terms of representative government, the church is dysfunctional. There's an us-and-them attitude toward the urban newcomers that's nonsense. So Crow Creek is just a bedroom town. Does that mean it's dying? You either adapt or you die in this world. If it's going to be a bedroom town, then it ought to be one. You see the Greens and Ed Biggs and Louise Fisher are adapting because they were never part of an elite power group like the Wests and the Pierces and the Scots. The tensions and anxieties are produced either by not letting go or trying to create something they can control. Most of the time in group dynamics, when you see this kind of symptoms, it's because people are afraid the group will die. It keeps other people from participating. All decisions are taken between these three families."

I tell him what Maggie Novak says, that you have to be a Methodist to really be part of Crow Creek. Rev. Bachman says she is probably right.

"Say you're an outsider, one of the 'bedroom people.' If it's very clear to you that your voice is irrelevant, that if you raise an opinion somebody doesn't like, you'll be yelled at, are you going to risk that? Absolutely not. In the old days Crow Creek was a little farm community. Every quarter section around Crow Creek had a farm or two. There were cows and chickens and pigs and everything else. It worked because everybody knew what they were supposed to do. You know, similar goals, similar needs, similar rules." Culture itself, as I said at the start, is a set of rules, a ready-made design for living handed down from parent to child, one generation to the next. This means the more traditional a society, the more rigid its rules and restraints are, the less freedom of individual choice there is.

"It worked," Rev. Bachman goes on, "because the choice to work in Cedar Rapids was not there. Choice was limited. Restrictions were there. Watch women here who assume the old rule. It's as if their husbands told them: 'Say what you want, but do as you're told.' It only worked back then because women had so few choices. You could leave this clod but you could only either be a nurse or a schoolteacher, if you had the training, or you'd starve. Even now, it's a human tendency to think you don't have choices. And as soon as you buy into that, you start to suffer, feel you're trapped."

What about the way, I ask, in such a small community, that everybody knows what everybody else is doing. Crow Creek's people are intensely interested in each other. I'd heard a lot of gossip. He agrees. "There's no place to talk that's safe. It's another way that Crow Creek is dysfunctional."

I gather he means that a lack of privacy is another way it is not working as it should. "Certainly you can't talk at the church," he goes on. "Because if you open your mouth, the enemy is going to say, 'Oh, I'll never forget that, partner.' And they'll crucify you next time they get a chance."

A lot of the Methodists, I argue, seem content, especially the older people who live downtown. He agrees. "People like Alice and Jonas and Ed are happy because they're not trying to preserve something that's dead. Alice and Jonas and Ed have nothing invested in the old days. They're just trying to live their life as best they can. They're not trying to preserve a culture.

"So what if farms are getting fewer and old farm communities die or get turned into bedroom towns?" Bachman asks, getting quite heated. "Is something so terribly valuable to American culture being lost? No. No, I don't think *anything* is being lost. I think the anxiety you find in Crow Creek is caused by people not understanding this. They want the picture to stay as it was in 1960. Or 1940. That's not possible."

But, I ask, what other base does our urban culture have? For instance, just take religion. It is the core of culture. Without some spiritual belief, any society loses its sense of purpose, its inventiveness, its industry, its enterprise, its faith in the future.

"The Methodist church is not the center of Crow Creek's culture," Rev. Bachman argues. "I don't think there is a center anymore. Religion might be. Don't confuse the two. There are large numbers of people in Crow Creek who have no relation at all to the church, any church. Okay? And why not? Both churches I serve are essentially secular. In my view, they are social phenomena only superficially tied to the Gospel. Who in Crow Creek really wants to discuss what Jesus taught. One of the tenets of the Christian faith is that your only purpose on earth is to serve the Kingdom. And that preserving such things as culture is irrelevant. They are only relevant in that they serve the Kingdom. To equate culture and religion together is blasphemy. Absolutely. It's one of the causes of neuroses."

I argue with him. He agrees with some points. "You can say we need to keep so many Americans on the land and in small communities if we're going to retain our ethics and our agricultural moral code. I won't argue with that. But you can't do it the way it was done in 1940. Because it's not 1940. My wife works in Cedar Rapids. She gets in the car and she's there in 30 minutes. Everybody in Martelle does that.

"I'll even agree that the Golden Rule at the center of our religion may come from farming, your idea that you help your neighbor get his crop in and he'll help you. But even that didn't fare so well in 1940 either. In 1948 there were thirty lynchings in Georgia. You would have met some wonderful, nice small-town and rural folks in Georgia. Wonderful and nice as long as you were white and southern and Protestant. Plenty of nice rural Germans had no trouble persecuting Jews. So for them the Golden Rule was a lie. It was a joke. And that's what you get when you confuse churchgoing with the Gospel, something that's upset theologians for centuries.

"Religion always has to fight the battle between the Gospel imperatives and how it is manifested in everyday life. That's what preaching is. It's calling people to account. You can't be a racial bigot and be a Christian. That's not acceptable. You can't persecute people and be a Christian. That's not acceptable. You can't oppress women and be a Christian. That's not acceptable. We have to confront that. They're contradictory. You can't come to church and just play church and be religious. That's heresy. It's blasphemy. It's hypocrisy. Jesus confronted that. I mean it's not a Crow Creek problem; that's a human problem."

It was like talking to a devil's advocate. To me, there is a degree of alienation to all urban life, regardless of a city's attractions. A sense of community, autonomy, and tradition is inherently stronger in rural communities. This is not a value judgment, one being good and one bad. Rural culture is simply the basis of all urban culture.

"You can talk about cities, New York, Chicago, pornography, drugs, crime," Rev. Bachman goes on. "Well, we have spouse beatings in Martelle. We have people who get drunk every night. If I had to rate films of naked people making love together, I find it less offensive than some of the incredible brutality I've witnessed almost every day in rural America. Some of my churchgoers get so righteous it frightens me."

Yes, I say, but there are also those who try to live by Christian ethics. On this he agrees. "There's those like Mary Pierce," he says. "She's caught in the middle. She at least knows what's going on. But she's trying to be everything for everybody. You can't do that. She thinks her job is to fix it. Ministers have the same problem. I do. I think I'm supposed to fix everything."

Bachman says that any time group dynamics are changed, there is stress. "But every time I hear somebody in Crow Creek say they want to keep the old traditions going, what that translates to for me is that they want to keep the power. They want to keep things as they were so they can run them." He was against the old handed-down rules and for freedom of individual choice. "My plan," he says, "for good or bad, is to preach the freedom of the Gospel. And that you don't have to be anywhere you don't want to be. You have freedom of choice. I'd like to push all these old walls so they'd have to collapse. And Crow Creek's people would be forced to deal with one another. Push these very thin walls. And clearly show in sermons and everywhere else the difference between how these people actually live and the radical demands of the Gospel. And let them deal with that however they will."

Not long after this talk, my visit ended. The last time I was in Crow Creek, Rev. Bachman was already at his new church near Des Moines. He has been replaced by an elderly retired minister, who just comes out on Sunday mornings to give a sermon; he sticks pretty close to the Bible. Crow Creek's walls, such as they were, did not get pushed, nor did they fall. The community remains divided between its city commuters and its beleaguered farm families.

On this last visit the talk was about the arrest of young Mike German, a Crow Creek boy whose widowed father is a security guard. When the mother was alive, the whole family came to the Methodist church. The father doted on Mike, and when he was a boy and had his heart set on having a white stallion, he got him one. After high school Mike went off and joined the Marines, but when he came back, he never seemed to settle down to anything. He lived at home but didn't seem to have a steady job. Then one night, when he was in his early 30s, he was arrested in Mount Hope Park, just down along the Wapsipinicon a couple of miles from Crow Creek. He was dealing in cocaine. On his person, or at home in his room, the sheriff found $130,000 in cash. He got a 7-year sentence in federal prison, reduced after he gave state's evidence. Some prominent businessmen and bankers in Cedar Rapids were involved; it was a whole ring.

Everybody in Crow Creek knows Mike. Adam still feels bad, as Mike used to drop by a lot to talk about cars and machinery and the Marines. At the trial, Maggie Novak tells me, one dealer testified that his drug clientele mainly lived in rural settings; they felt safer if they moved out to small farming communities.

Now in a fourth and final section we will see how the Prairies and Crow Creeks might be saved at a time of grim predictions that half of America's farms may vanish in the next 15 years.

# IV

## Going, Going

M Y MAIN ARGUMENT in this book, if you'll accept that I'm painting with a very broad brush, is that what kind of urban culture we have in America is going to depend on how many Americans farm.

All culture, as we've said, has a rural origin. To farm is to till the soil. It needs land (property), tillers (family), and a set of rules. Indeed, the Latin *agri* for land and *cultura* for cultivate is how we got *agriculture*. In the late twentieth century we find some breakdown in property, family, and rules in our American cities. Successive generations are growing up without rural ties. This in turn indirectly produces social ills like homelessness, crime, vandalism, and drug abuse. I mean this not in a sense of urban-bad, rural-good. Nor as a romantic pastoral view.

Rural life can have its shortcomings. Farming has always meant hard physical labor. Socially, as Richard Lingeman put it in his 1980 work, *Small Town America,* a little community can be "good, generous, kind, helpful in trouble, cradle to grave." But it can also be "materialistic, insular, suspicious, set in its way, canny, backbiting, smothering. . . ." To live in a narrow world is to risk being narrow-minded. But my main point is: farming creates societies that work. It creates a very durable and basic culture. And we need to save as many farms and small towns as we can because America's urban culture is at stake.

No society can get too far away from its rural origins and farming and stay healthy. At bottom, this is what ails Britain, the first country to industrialize and urbanize. Japan, the other most urban major country, where industry came later, has somehow managed—with its post-Confucianism, closer rural ties, and village-type urban social organization—to do better so far. The Third World, made up of predominantly

203

rural countries, has a fragile urban stability that, I think, depends on the still-rural roots of most of its city dwellers. I dread to think what will happen as successive urban generations lose their village ties on a scale anything like we are experiencing.

In this fourth and final section, I'll first look at how the declines in farming and family are related, how the 1980s compare with the 1930s, and why the small town is being much harder hit this time. Urban culture, I repeat, is the issue; at its heart is the family. For urban culture to stay healthy, farms have to survive and people have to farm. This requires wise economics. So I posed the question to a number of experts: how do we solve the American farm problem?

They are authorities in their fields whose ideas I have come to value over the years: Don Paarlberg, former chief economist, U.S. Department of Agriculture (USDA); economist Lowell S. Hardin, who formerly ran the Ford Foundation's agricultural programs; economist Neil Harl, Iowa State University; Wayne D. Rasmussen, veteran chief historian, USDA; historian Gilbert C. Fite; Norman E. Borlaug, plant breeder and recipient in 1970 of the Nobel Peace Prize as father of the Green Revolution; the University of Chicago's Theodore W. Schultz, as mentioned, a 1979 Nobel laureate in economic science; Calvin Beale, chief demographer, USDA. Other opinions are also given.

After the farm problem, we turn to the universal rural culture, what is distinctive about it as found in Third World villages, and how much of it appears in Prairie and Crow Creek. At the close we hear from William H. McNeill, professor emeritus of history at the University of Chicago and author of *The Rise of the West*, to many, as noted, the best world history written by an American. Though, on their face, our backgrounds of study and experience could not be more different, on the question of what is happening to our rural and urban culture, we come together on significant points. I have kept Professor McNeill's views for the last because I feel that what he has to say is fundamental and perhaps seminal.

Scholars on the American family agree that the ideal of the breadwinner-homemaker couple thrived best between 1860 and 1920. "Only connect," E. M. Forster said. These 60 years were also the heyday of the family farm. In 1860 there were 2,044,077 farms, nearly 600,000 more than in 1850, as railroads and settlers equipped with John Deere's sod-breaking steel plow surged westward to the Pacific. The rural population—officially deemed those on farms or in towns of less than 2,500—grew from 18 million in 1860 to over 50 million by 1920, when there were about 6.5

million farms. (As I mentioned when I gave these figures earlier, they peaked in 1935 at 6.8 million.)

Cities also grew, and much faster. New York went from 1.2 million in 1860 to over 5 million by 1920. Chicago went from just 109,260 in 1860 to 2.2 million by 1920, mainly as a manufacturing and marketing center for farm machinery and produce ("Hog Butcher for the World/Tool Maker, Stacker of Wheat"). More than 75 percent of Americans lived on farms or in rural communities in 1860; 30 percent still did in 1920.

The same years, 1860–1920, culturally saw the decline of Victorian values (Queen Victoria died in 1901), such as the father's absolute authority and control of money (read Clarence Day's *Life with Father*). Better education for women and the Crash of 1893, forcing more women to work to eke out family incomes, also weakened the urban family. There was a slow, steady climb in the rates of working women and divorce. And then, just after World War II the family—and farming—had a respite as soldiers came home, fertility rates rose, and we got the famous baby boom. By 1956 the median age for marriage fell to 22.5 years for men, 20.1 for women; that year there were still 4.5 million farms.

Then in 1960–70 the number of farms fell by 1 million to 2.9 million. In this same decade the fertility rate went way down and the age of first marriage and the divorce rates went way up. More women worked. Nearly 20 percent of mothers with children had jobs; nearly 40 percent of those with children between 6 and 17 had jobs.

Even more telling, whereas six out of every ten Americans made their living farming in 1860, this was down to one out of ten by 1956 (and to two out of a hundred now). The peak farming year of 1935, when a fourth of Americans farmed, is as remote to young people today as the Civil War was to the Jazz Age, or the Jazz Age to today's youth (5 percent of them asked by the Iowa Poll in 1989 said they were personally affected by the death of British punk rocker Sid Vicious; but don't give up hope on Iowa—87 percent still say it's not the heat, it's the humidity).

These 60 years, 1860–1920, besides being the best years of the family farm and of the American family itself, were also the heyday of the horse. The 3-mile-per-hour gait of a horse pulling a plow, reaper, or wagon set the rhythm of rural life and meant the average farm was about 160 acres. Even by 1940 this was only up to 175 acres; it is, nationally, 440 acres today. The railroads, as I mentioned, when they set up towns across the Great Plains, felt they were ideally 6 miles apart, so nobody had to go farther by horse and wagon than 3 miles, an hour's trip. Of course by

1940, horses were quite suddenly replaced by oil-fueled machines. The car, truck, pickup, and rubber-tire tractor made it possible for farmers to travel easily and quickly to distant towns which were larger and had more goods and services.

So that a great many little towns, created by the railroads in the 1880s with the end of the frontier, by 1940, 60 years later, lost their reason for being when horses were replaced by cars and machines. Amazingly, people being tenacious once they have sunk their roots, many are still hanging on. But small towns have been worse hit in the 1980s and 1990s than they were in the 1920s and 1930s when horses went out. Are there parallels between the Great Depression years and now? As we saw, people in Prairie and Crow Creek think so.

Agriculture did go through a 1970s-style land boom during World War I. Farm prices shot up. My father, for example, was farming in North Dakota's Red River Valley in 1916–17 when potatoes went from 80 cents a bushel to $2.50 in a year. Wheat prices soared. He made a killing on hogs and lambs. It allowed him in 1917 to go back to medical school.

Then commodity prices sharply dropped again in 1919–21. Land values plummeted. Farming bounced through the 1920s, not a very good decade. The 1929 stock market crash and bank failures wiped out savings and risk capital worldwide. Commodity prices took another sharp drop in 1931, probably because of overproduction. There was a total collapse of liquidity in rural areas. My father, by now a country doctor in Prairie, had to move his wife and five children to Fargo, as in Hawk County almost nobody could pay their bills. Unemployment, low grain and livestock prices, drought, and grasshoppers destroyed the farm economy for years.

Conditions in 1987–90 have not been this bad. In the 1970s American farming had its biggest postwar boom. Exports to Europe, Russia, and the Third World, helped by a devalued dollar, shot up. But land values inflated, as they did in 1917–19, and left farmers with heavy debts when the 1980s brought falling prices, shrinking markets, and rising farm surpluses. Farm after farm went broke; there were heavy debt-to-asset ratios.

Then came the 1988 drought. Prairie's farmers, as we've seen, lost about two-thirds of their crop, Crow Creek's farmers about half of theirs. Grain and livestock prices were good. Diversified farms and those with big inventories of grain, like our best Prairie farmers, Nels Peterson, Paul Fischer, and Harry Schumacher, did all right. With crop insurance and disaster relief, and by selling stored grain, most farmers pulled through. But 1989 was again a very dry summer. Grain prices went down, most farmers had already sold what was in storage, and crop insurance and disaster payments were less than in 1988, an election year. The 1989–90 winter had little snow, and in eastern North Dakota, a plague of grasshoppers was another echo of the 1930s. The hard times are not yet over. So far we have not returned to the upsurge of bankruptcies and auctions of 1987, but as I said at the start, it will take several good years yet to get out of the woods.

Remember Bud Petrescu quoting his father that during the Depression and Dust Bowl years if a farm family had a few cows, hogs, and chickens, even if it lost the crop, it lived and kept the farm? America's rural population stayed pretty much the same in 1925–35. People raised big gardens and canned. They survived on little.

It is the collapse of small towns now that is worse than anything 60 years ago. Lowell Norland, a former farmer and politician who was doing community work with the University of Northern Iowa in Cedar Falls when I interviewed him, said, "A week ago Sunday I was home. The grocery store's gone, the drugstore's gone, the hardware store's gone. The school is going. The church is still there, but it has to share its preacher with two other congregations. People are cracking. There's psychological stress too. The stress is absolutely enormous out there. I went to the opening of a new shopping mall in Mason City. A friend who'd been a farmer was going around in a uniform. He'd inherited his farm and was a hard-working man and still he lost it. He was doing custodial work. He was the custodian who was going around picking up cigarette butts."

Norland, who was the Democratic majority leader in Iowa's House of Representatives for 14 years, said that as times get tougher, it is tougher still on the minorities. I talked to a black shoe shine man in Waterloo. For years he had worked at the huge John Deere tractor factory, at last getting laid off. "I made $12.50 an hour. Now maybe I make $300 a month. Can't hardly live on that. Some days nobody comes for a shine at all. I put my trust in Jesus. In these hard times I put my trust in the Lord. Ain't nobody else you *can* put your trust in."

Norland himself took over his father's farm in 1951. "Farming was my main occupation for 20 years. When I was home, it used to be I knew every farm, how many children the family had, what church they went to. We've lost that stability. To me that's the biggest change in my lifetime.

"There was an awful lot of growth in the 1960s and 1970s. People came in and bid up the land and the farms got bigger and bigger and bigger. And the farmers got fewer and fewer and fewer. The grocery store left, the drugstore left, the hardware store is going. What you have left is a branch bank, a quick stop like Casey's, a filling station, and not very much else. In my town the school is gone. The church is still there but the preacher has three or four congregations. And you've got instability of the family, the community, the whole culture."

Economist Kenneth Stone at Iowa State University told me small towns of less than five thousand in Iowa lost a third of their retail trade to larger towns and cities in the 10 years, 1977–86. Between 1979 and 1986, Iowa lost a fourth of its grocery stores—455 of them—5 percent of its gas stations, 6 percent of its variety stores, and 6 percent of its boys' and men's clothing stores. Farm businesses were badly hit; 7 percent of building material dealers went out in 1986 alone; 5 percent of all hardware stores closed that year. Go around Iowa and you find old implement dealerships empty, hospitals just a third full. Taverns are growing in number, if total sales are off. Rural Iowa's one booming business is nursing homes; most have long waiting lists.

Remember Adam Pierce saying that "when Wal-Mart came in, it killed Main Street in Anamosa"? The 1,400-store Wal-Mart chain of giant discount stores increased its annual sales to over $29 billion the past decade, mainly by setting up shop all over rural America. Its owner, Sam Walton, is now said to be the world's fifth richest man. But every time another Wal-Mart opens, underselling everybody else, Main Streets for miles around are often converted into wastelands of empty shops with broken windows. Professor Stone, in a 1989 study, while protesting, "I believe in capitalism," found many local businesses cannot compete with such mass merchandising.

In Iowa, small towns don't die, their Main Streets do. Residential areas, as in Crow Creek, fill up with urban commuters. Even so, younger people are moving out of the state. Iowa lost 2.5 percent of its people in 1980–89, dropping from 2,914,000 to 2,840,000. Outside of the District of Columbia, also down, the only other state to actually lose population was chronically depressed West Virginia. States that were the biggest gainers the past decade were Nevada (up 38.8 percent), Arizona (30.8 percent), Florida (30 percent), California (22.8 percent), New Hampshire (20.2 percent), and Texas (19.4 percent).

North Dakota just held its own, gaining 1.1 percent, though there was that huge drop in 1984–87 of close to 15,000 people, a real hemorrhage in a state with only 660,000 to start with. Between 60 and 75 percent of the graduates of North Dakota's two universities leave the state for greener pastures. North Dakota's only fast-growing population, as I mentioned, is the Native Americans; its Indians grew by 50 percent in the 1970s and 1980s. In terms of jobs, the Federal Bureau of Labor Statistics reported in 1990 that Iowa ranked forty-sixth nationally in growth employment from 1987 to 1989, North Dakota forty-second. Florida and Arizona were the top two.

In the 1970s, when the value of farmland boomed, as noted, North Dakota had America's highest percentage of millionaires of any state. Today it is 15 percent below the U.S. income average. North Dakota doesn't have to worry about Wal-Marts; its little towns have to worry about staying alive. It is now down to 90 small towns of 500 to 2,500 people, and about 250 more with less than 500 people. North Dakota's number of farms was down to 33,000 in 1990, with average acreages up the past decade from 1,000 to over 1,200. The value of all crops except potatoes and sugar beets fell in the 1980s. North Dakota is sitting on more than 500 billion tons of lignite coal, but, as with oil, coal prices have to go so high before it is economical to mine. There are less than 15,000 manufacturing jobs in the whole state.

If you like empty space, North Dakota has real advantages. It has America's cleanest air and lowest infant mortality. It has the country's third lowest crime rate and is the only place I know left where people in places like Prairie often do not lock their doors at night.

Those 90 500-to-2,500-people towns badly need the help of foundations, think tanks, or people with ideas on how they can attract "value-added" service industry. This would provide jobs and let many go on living on the land and farming part-time. Towns like Prairie need outside help because if small service industries are going to survive in today's single-world economy, they have to be competitive, be able to adapt to rapidly changing technology, keep production flexible, sell complex goods to sophisticated customers in short-lived markets, and emphasize quality over quantity. This is way beyond the ability of most of the country's Prairies; outside skill is needed.

Otherwise, for Prairie and the other 89 small North Dakota towns like it, the future looks bleak. Charlie Pritchett said Prairie is getting close to the point where there isn't enough in it "to bring people to town for *anything*." This has already happened to Sheridan County, just west of Hawk County. It no longer has a doctor, hospital, traffic light, or Greyhound bus service (neither does Prairie anymore, just a funny little once-a-day minibus). McCluskey, Sheridan County's biggest town, has barely

600 people, to Prairie's 700 or so. McCluskey's Main Street, like Prairie's, is half made up of storefronts that have stood empty for years. Calvin Beale says that every decade, when the census is taken, about 200 such previously incorporated towns across America fail to show up.

In December 1989, North Dakota repealed a tax increase intended to offset a $100 billion budget deficit. It has to make across-the-board cuts in all its services. Outside experts urge it to consolidate some of its fifty-three county governments and 296 school districts. But who wants their county seat or school to go? For Prairie it would be the end.

North Dakota's Governor George Sinner says the state wants to keep every community it can. "But when you have a declining population and tax base and have to keep schools going and the highways going and the medical services and the nursing homes, it's going to be almost impossible to keep these communities alive. The state just can't do it."

North Dakotans are divided about what to do. William C. Davis, the state's director of industrial development, says that "towns that don't have much of a trade area and don't have any primary government service will be the first to go." Others argue every small town has to decide its own fate. I'm with them.

Professor Mark Lapping, director of Kansas State University's Kansas Center for Rural Initiatives, has come up with a triage system. In battle, the wounded most likely to live get treated first. You need it in battle; I saw it in Vietnam. But it is of dubious merit when applied to other things. For instance, it had a vogue in development circles in the 1970s; the incurably ill—India was deemed one of them—should be cut off from foreign aid. Well, India fooled everybody and now feeds itself. So much for triage.

Professor Lapping's criteria for which small towns should live or die are size—he says over 2,500 people—economic diversity, outside corporate investment, the willingness of residents to invest there, easy accessibility, and a relatively young population. Lapping's view is that of an urban economist. I would argue, from a more anthropological viewpoint, that it is communities *below* 2,500 we should be helping. The smaller, the more rural, the closer organically to farmers and the land, the culturally better off all of us will be.

A recent study of ten deeply distressed North Dakota counties by Frank Popper, head of the urban studies department at Rutgers University, and his wife, Deborah Popper, found each had no more than four people per square mile, poverty rates of at least 20 percent, and new construction of less than $50 per capita (national average: $850). Prairie's Hawk County is one of the eighteen North Dakota counties that lost 50 percent or more of its people between 1930 and 1988.

Some federal farm initiatives, like the Conservation Reserve Program (CRP), which pays farmers to take land out of production, hurt small towns. In Hawk County the CRP has caused a drop in demands for local goods and services. There are no consumers on empty acres.

So far North Dakota is giving more help to towns that help themselves. "That's just the way the world works," says Governor Sinner. "There's a limit to what we can do." The state offers low-interest loans and matching funds to towns that "take the initiative" by raising money of their own to attract businesses. For example, Park River, a town of two thousand north-west of Prairie, spent tens of thousands of dollars to improve a site and provide low-cost power for a high-technology plant to harvest fresh mush-rooms. In late 1989 the Midwest Mushroom Company began marketing mushrooms to wholesale distributors in four states. Iowa State's Neil Harl urges caution about such efforts. He says one mushroom company is now in bankruptcy and "there are a lot of unhappy people as a result."

Calvin Beale, as a demographer, suspects that outside the Middle West, which now has 51 percent of what farmers are left, a lot of the potential rural decline in America has already taken place. He expects the number of rural Midwesterners to continue to fall wherever there is a lack of timber, minerals, or attractions for retirees or vacationers.

Beale says places like Crow Creek, which demographically survive by going from a farming market center to a residence for commuters from the city, have a real identity crisis. "When is a town a town?" he asks. "Is it any longer meaningfully such if there is no longer a school, or bank, or drug store? In the past, as population loss occurred, hundreds of counties and towns had a cushion of people and business and institutions within which to absorb some loss without a radical change in the character of the place. Now, after a generation or more of decline, they are falling through one minimum floor after another of population size necessary to support life. And the Wal-Marts, or even regional shopping malls, within a half-hour or 45-minute drive, make it worse."

Beale notices, when farms and small towns go, there is a big drop in self-employment. "The smaller the scale of the place, the higher the average of people who are self-employed, which I think results in greater stability and personal satisfaction." In 1980 self-employment in cities was just over 5 percent, but in towns of 2,500 to 10,000 it was 7.3 percent; in towns of 1,000 to 2,000, 8.6 percent; and among people on farms, 38 percent (which means there were a lot of wives with nonfarm jobs and a lot of part-time farmers).

Beale is also concerned that the Middle West could, if farms get big enough, go back to hired men and become like California, where most farm work is done by wage labor. This would be a further erosion of the family farm. Ernest Waldenberg's two hired hands are our example of this.

As the thousands of little towns go downhill, their people, as we saw in Prairie, are wishful and resigned, caught up in forces far beyond their control, and determined to hold on as long as they can. They do not feel culturally threatened as Crow Creek's native-born farmers do, resenting the urban invaders and nostalgic for the recent past.

My guess is that what farms and farm marketing communities survive, if fewer in number, will be as culturally healthy as they are today. Wayne Rasmussen, asked what he thought a typical American farm would be like in, say, 2020, didn't see it changing much. "First," he said, "it will be a large commercial enterprise operated by a family. Second, the internal combustion engine will still be the primary source of power unless solar energy is harnessed by then." He did see computer-integrated automation further reducing farm drudgery and the farm family enjoying "virtually all the amenities of the city dweller." But crucially, it seems to me, Rasmussen predicted this future typical farm family will still see themselves as "being a group apart that works with the soil, water, and seed to provide large numbers of other people with a plentiful supply of food." A family of tillers cultivating their property of land—the essential cultural base stays intact.

Farmers like Clint Houseman in Crow Creek or Bud Petrescu, Paul Fischer, and Harry Schumacher in Prairie are practically there now. With their university educations, two-way radios, computers, fax machines, and electronic market services, with their digital-controlled and automated machinery, and homes as comfortable as those in cities, they are well into this future already, even as their way and view of life remain emphatically rural.

As I keep repeating, the fundamental issue here is urban—what kind of a culture is America going to have in its cities? But if the ultimate issue is urban, the place to solve it is on the farm.

NOW LET US turn to the experts. First, how do we define the farm problem? I asked Don Paarlberg. "If the problem is to restore the family farm as grandfather knew it," he said, "there is no solution. If the problem is to generate a satisfactory income for everyone who wants to farm, there is no solution. If the problem is to provide an adequate income for efficient farmers and a healthy rural economy, there *is* a solution. But the old order is passing. We can't put the chicken back in the egg."

Paarlberg's solution: "An open system for productive agriculture and off-farm job opportunities for those who, for whatever reason, cannot make it as full-time farmers."

Economist Lowell S. Hardin, who observed I left it wide open to everybody to define the farm problem for themselves, ends up pretty much in the same place. He would make rural America an ever-more-attractive place to live. "This requires zoning to protect property values and enforced rules of the game with respect to the environment. It calls for improved infrastructure (schools, roads, utilities, health services, and protection of the environment), especially in more populous areas.

"The goal is to enable and encourage more people to live in the country-side as rural residents and part-time farmers while earning most of their income from off-farm sources. As I see it, this is how one preserves the much-cherished small farm. The commercial farmer, who also benefits from the improved infrastructure, in turn becomes increasingly reliant on the marketplace." But Hardin warns it will take education, training, and jobs for poor farmers, not acreage allotments and deficiency payments, to enable them to make the adjustment. (I've known Dr. Hardin, like Paarlberg, for years, first interviewing him in 1968 when I was doing a profile of his cousin, Secretary of Agriculture Clifford Hardin, for *The Washington Star*.)

To Iowa State University's Neil Harl the cost of capital and labor is key. "In the 1970s," he told me in 1987 in Ames, "capital was viewed as exceedingly cheap. Its real cost—the amount of interest people pay at the bank less the rate of inflation—got as low as 1 percent. And labor was perceived as high. So you saw substitution of capital for labor. Big four-wheel-drive tractors. Big combines. You saw bigness and concentration."

Today Dr. Harl, an extremely active economist, who is also engaged in Poland, proposes this solution: "Get the real cost of capital up and the cost of labor down as in the 1980s so you can see a substitution of cheaper labor for expensive capital; people take more time, use smaller equipment. This is more friendly to family farms. And less friendly toward expansion and bigness. The problem with that is the eternal quest for higher incomes makes that outcome very unlikely. Short of a ban on manufacture of large-scale equipment or a tax on large tractors or combines, there's little chance of this happening. Even the current emphasis on limiting chemical use is likely to result in only a limited increase in more labor-intensive methods of weed control including mechanical cultivation."

Wayne Rasmussen, longtime chief USDA historian, says, "I am afraid that the farm problem will be solved in the future the way it is now being solved: a continuing consolidation of farms into larger units, with greater applications of technology in their operations." Rasmussen predicts there will be fewer full-time farmers or family farms, but more part-time small farms—rural residences with some farming. I suppose Adam Pierce would qualify for this, as would have Charlie Pritchett when he both farmed and edited his newspaper. But central North Dakota offers so few nonfarming jobs. Rasmussen feels that in such sparsely populated areas—he himself is from eastern Montana—rural communities will continue to decline.

In 1925, Rasmussen says, his native Golden Valley County in Montana had 492 farms of an average size of 637 acres. In 1979 there were just 139 farms averaging 4,693 acres. Like Hawk County, Golden Valley County lost half its people in 50 years, 1930–80; today it has just over a thousand. Whereas in the 1920s it had two banks, two newspapers, two small hospitals, three doctors, a railroad line, and a creamery, by 1980 they were all gone. One of three high schools has been closed; the number of elementary schools has dropped from twenty-five to two. The roads, Rasmussen says, are better than ever, but young people are leaving faster now than in the 1920s and 1930s.

His solution to the farm problem: "We need a more rational system of farm subsidies. We need some way to insure a sure, safe supply of food at reasonable prices, but we do not need to subsidize the production of every bushel of wheat or every pound of tobacco as we have been doing since the present price support system was first adopted in 1933.

"Instead, I would wipe out all commodity price supports, acreage controls, export subsidies, tariffs and import controls on farm commodities, and related aids to farmers and agribusiness. I would encourage our farmers to compete on the world market, and I am convinced that they could do so in such commodities as wheat, corn, soybeans, and rice. Some producers of some commodities, such as sugar, would be forced to move to the production of other commodities.

"However, I would guarantee every family operating a farm from which they usually derived most of their income an income comparable to that of the average city dweller, providing they did a reasonable job of operating the farm. I would insure against natural disaster with crop insurance or disaster relief of some type.

"The easiest and probably fairest way to administer the subsidy program would be through a 'reverse income tax,' an idea that has been advocated by others. Every full-time farm family would be eligible, including families operating very large farms, on exactly the same terms. Large corporation farms would not be eligible, nor would investors in farmland. However, it might be reasonable to make some special allowance for small or part-time farmers."

Wayne Rasmussen feels such a system would encourage more people to stay on farms and might lead to the breakup of very large farms. "Will some such system as the one I propose be adopted?" he asks. "Of course not. The best we can really hope to do is to encourage the trend towards small and part-time farms supported by off-farm employment and thus maintain the viability of some of our rural communities that otherwise will disappear."

Historian Gilbert C. Fite, whose 1981 book, *American Farmers: The New Minority,* charts rural decline, says that since America is a democracy, it pretty much rules out radical solutions. I agree. We can't impose ceilings on land or on the size of machines, though either would save lots of family farms and small towns. And we still don't really know what genetic engineering, as Harry Schumacher put it, "an avalanche" bearing down on us, will do.

Dr. Fite's solution: "I would continue target prices for the basic grain crops of corn, wheat, rice, and soybeans, and some modest supports for cotton and milk. Because of the farmer's poor bargaining position with other elements of the economy, I think that some price supports are essential for farm welfare. I also favor acreage restrictions in return for price supports in order to keep large price-depressing surpluses from occurring. Because of subsidized farm production in other countries, I believe we should also have a two-price system—a higher price for domestic sales and a lower, world-competitive price for exports.

"My national farm policy would also contain a very hard-fisted position with countries like Japan to open their markets. I believe that if we told the Japanese that we would cut automobile imports by one-half until they opened their agricultural markets, that we could get some action. Moreover, I think farm policy should move faster and further toward converting grain to fuel, both to use up surplus corn in particular and to help clean up the environment."

Earlier I quoted University of Minnesota economist Vernon Ruttan's observation that the biggest gains in American farm production have come in oil-based mechanical technology, or advances in output per worker, while the biggest gains in the Third World are coming in biological technology, or advances in output per unit of land. Despite huge crop increases, farms in India and China are not getting bigger and fewer the way they did in America. (Indeed, the number of draft animals in both countries has grown.) Biological technology (by that I mean seeds, irrigation, pesticides, and fertilizer), unlike mechanical technology, does not demand the same substitution of capital for labor.

In 1990 I asked Professor Ruttan about his distinction between mechanical and biological technology, something I have been quoting for years. He said he still found it useful and that the substitution of machinery for labor in American farming was in response to rising wage rates in the towns and cities. If farm earnings were to keep up with urban wages, each farmer had to till more acres. "Mechanization," Ruttan said, "was the response to labor being pulled out of agriculture rather than being pushed out."

The present situation, he said, is hard to interpret. "As you know, wage rates in manufacturing in the United States have not risen for more than a decade and a half. My sense is that this should reduce the pressure for further substitution of machines for labor. However, this may have very few implications, since there is hardly anybody out there working. When you drive through the countryside in India, you find the fields full of people. When you drive from Minneapolis to Des Moines, you see very few men or machines in the field. Furthermore, the cost of adding additional horsepower to save an hour of labor has now become very high." Even so, he does not expect to see a reversal in land-labor ratios.

When I asked how he would solve the farm problem, he said, "First, the farm problem does not need a solution. The farm depression of the early and mid-1980s was due to the Volker-Reagan depression. It was not due to technology. Productivity growth both in the U.S. and in developing nations has slowed down during the 1980s. It is time to decouple income support from price support." Dr. Ruttan did agree there was a serious rural problem. He said, "Adjustments in the small towns and villages have lagged behind the adjustments in agriculture." He felt American farming could no longer support the old rural infrastructure, such as schools, hospitals, and roads, just as Governor Sinner said.

My guess is advances in molecular biology, like the great new breakthroughs in radio astronomy and solid-state physics, and unlocking the secrets of the atom, as well as such technological feats as rockets and spaceships to escape the earth's gravity or transistor radios and digital computers, will radically affect both urban and rural life. By *biological technology,* which Professor Ruttan compared with mechanical technology, I mean conventional plant breeding and the application of biology and chemistry to farming. *Biotechnology,* a related term, has come to specifically mean genetic engineering.

A big, still undecided question is whether its biggest gains will come in Professor Ruttan's output per worker or in output per unit of land. *Biotechnology,* like *global marketplace,* is a current buzzword. But its early delivery in new drugs and new kinds of fertilizer and pesticides, with enhanced livestock following, is unlikely to be followed soon by radically improved food crops. There are also unanswered questions about biotechnology's impact on the environment; some fear it will raise unforeseen problems.

Norman Borlaug's solution to the farm problem: "Keep making scientific and technological advances to keep America's lead in science-based biological, chemical, and mechanical technologies. Lower land values and lower-cost production techniques are also needed to keep America competitive."

Dr. Borlaug forecasts continued progress in plant breeding, cloning, and tissue culture but is cautious when it comes to biotechnology in plants. "I'm pretty sure genetic engineering or the transfer techniques at molecular level will produce better vaccines. There's been a lot of progress with bacteria and yeasts to produce insulin, interferon, and growth hormones. I think, too, there's a case in genetic engineering to combine several genes into a broader spectrum of antibiotics.

"But when it comes to higher plants, these are pretty complex things. Leaving asexually propagated species such as potatoes aside, there is little evidence as yet that genetic techniques can produce higher-yielding crops with greater disease and insect resistance in the next 10 to 15 years. Far less basic research has been done in the role of genes in the higher plant species than for humans or animals with their complex immune system, largely as the result of cancer research."

Ever since two scientists at Cambridge University, American James Watson and Englishman Francis Crick, worked out in 1953 the double-helical structure of DNA, the chemical blueprint of all living creatures, biotechnology has begun to transform farming. The British in 1986 put DNA in a single cell of rice; the Japanese have now gone further, getting whole plants. But, as Dr. Borlaug says, genetic engineering in tubers, trees, and ornamental plants has not been matched in grasses like wheat, barley, rice, and corn. The hope is someday gene implantation will make plants resistant to insects and viruses, heat and cold, floods and drought, and speed photosynthesis, growth, and nitrogen fixation.

But already embryo implants enable cows to have five or six times more calves (up to eighteen), and growth hormones increase their milk yields. Genetic engineering can already lower fat content in beef, choose the preferred sex in offspring (females for dairy cattle, males for sheep and beef cattle), treat animal diseases like diarrhea, fever, and cancer, and improve vaccines. Our growing knowledge of molecular biology is already put in use to ferment wine and cheese, create artificial sweeteners, and alter the calories and flavors in foods. Soil microbes now treat chemical wastes; others control fungi.

As I said at the outset, science, in its practical application as technology, is the agent of change in a rural culture, something I observed again and again in Third World villages. Take the American dairy industry, for instance. In 1989–90 it was booming. Drought and forage problems in mid-1989, together with an unforeseen drop in foreign milk production, produced a surge in demand for American dairy products. Operating margins—the difference between costs and income after depreciation—were $322 per cow in 1987, fell to $134 in 1988, then rebounded to $253 in 1989, in some areas to nearly $600 per cow. Nobody expects this to be a long-lasting resolution.

The reason: biotechnology. The number of dairy farms in America fell by 40 percent between 1978 and 1988; there are now just 220,880. But, with genetic engineering, milk per cow is expected to go up at least 50 percent the first year and eventually to 100 percent. Which means 100,000 dairy farms could produce today's output. This has become a hot political issue in Wisconsin, with the state divided between those who would ban genetically engineered drugs to increase milk production, and those who see biotechnology as its rural people's biggest hope.

The Office of Technology Assessment in 1986 forecast that biotechnology in crops would be more quickly adopted by richer farmers who could afford it, leading to still bigger and fewer farms. Others argue that the more that gets built into the seed itself, the more it means higher yields at lower cost. If this happens, it could drive down land values and the prices of fertilizer, pesticide, and other inputs. If it reduced farm income, it could work to the small farmer's advantage. As it is with all new technology, it is hard to foresee the consequences.

Some political activists, like Ron Krupicka, head of Small Farmers

Resources Project in Hartington, Nebraska, argue that land ceilings could be imposed to prevent big farmers from using "bigger and better technologies that could harm the land resource." His solution: raise farm subsidies and limit the size of farms. I'm with him but most farmers aren't. Propose land reform to the average Midwestern farmer and he'll call for the sheriff. "Land to the tiller!" is the populist battle cry in much of the Third World—and paradoxically, now in Russia itself—but not in America, even in places like Hawk County, where tenant farming runs 60 to 70 percent and there are high rates of absentee landlordism.

Not only does no constituency yet exist either for redistributing land or putting limits on its acquisition, but American agriculture is losing what little constituency it has, both in terms of farm voters and congressmen. Don Paarlberg says that whereas 251 of 435 congressional districts were "farm districts" in 1926—meaning that at least 20 percent of their people lived on farms—this dropped to 49 by 1976. Farmers are no longer the biggest single occupational group in any state, not even one so heavily rural as North Dakota.

To some, Congress ought to pass stronger legislation to protect the small family farm. Rev. David Ostendorf, who headed Prairiefire Rural Action, Inc., in Des Moines when I interviewed him, said, "Ultimately the whole social fabric of rural America is at stake." Ostendorf's solution: get passage of a family farm act to provide a stronger farm management program, higher price supports, and farmer referendums so they themselves can vote on farm programs.

Environmentalists are concerned about carbon dioxide and ozone buildup, soil erosion, acid rain, groundwater contamination, and pesticide residues in food. Kenneth A. Cook of the Center for Resource Economics in Washington expects farmers in time to respond to ecological, economic, or regulatory signals and change their practices "to reduce the generation of pollutants and conserve soil, water, and wildlife." His solution: reduce oil-based fertilizer and pesticide by 30 to 50 percent in some areas and introduce farming practices favorable to small farmers.

Author Richard Rhodes blames chemically intensive farming on subsidies. To compete, he says, farmers must massively apply nitrate fertilizer and chemical herbicides. The farm of Tom Bauer—not his real name—which Rhodes focused on in his 1989 book, *Farm,* grossed $152,090 in 1986. He gives a breakdown:

"Their largest single expense was feed for their cattle and hogs. But fertilizer and chemicals took second place at $22,345. Add $17,910 more for machinery and fuel, and the Bauers' purchases from agribusiness are the biggest overall expenditure. In exchange for idling about 10 percent of their land, they earned some $11,000 in government subsidies. Agribusiness pocketed all those dollars and $29,000 of their unsubsidized gross income besides. The Bauers cleared only about $19,000 for their year's labor." Rhodes' solution: limit production of surpluses, encourage soil and water conservation, and begin moving farming toward less dependence on chemicals and more reliance on mechanical cultivation and biological controls.

Another proponent to do away with subsidies is Dennis Avery, former senior agricultural analyst for the State Department and now a Washington consultant. Avery argues the world market for American food and fiber is drying up. In 1988 India was able to cover a 20-million-ton grain shortfall from its own stores (and, according to Dr. Borlaug, its record 1989 grain crop of 54.4 million tons will be about 56 to 57 million tons in 1990, a huge rise compared with the 11 million tons a year India used to harvest in the 1960s). Then, Avery says, China exported more cotton and corn than it imported wheat in 1989. Brazil, another giant, Avery reports, has come up with a new high-yield corn bred for the acid soil of its remote Cerrado plateau; a Japanese-financed railroad is to be built to open up this vast area to wheat, corn, and soybeans. Avery's solution: do away with farm subsidies that protect high land values and the incomes of the biggest commercial farms in favor of free trade.

In 1960–61, our farm exports rose more than ninefold. But in 1981–86 the loss in dollar terms was about 40 percent. The lost sales were blamed on everything from an overvalued dollar to higher American production costs, new European trade barriers, and global recession in the 1980s. I think we must also face the new reality that the world has learned to use our science to grow its own food.

I asked Calvin Beale how he would solve the farm problem. Beale said there probably is no solution. "No advanced nation seems to have been able to stem the tide of consolidation. So long as productivity increases— whether through machinery, chemistry, genetic improvements, or biotechnology—I think the trend to larger commercial operations will go on. Only the part-time farmer will be able to remain with a modest-sized place."

The 1992 agricultural census—the last two were in 1982 and 1987— will also give us a clearer picture of American farming. Those in the 1980s revealed that a very large number of very small farms now produce just a small proportion of our food. By 1982, 60 percent of all American farms had sales below $20,000; they made up just 5.5 percent of all farm sales. But a mere 13.5 percent of all farms, with sales above $100,000, accounted for 75 percent of all sales. It looks like there may be only about 300,000 or 400,000 farms worth the name left.

The small farm of under 100 acres is typically like Adam Pierce's in Crow Creek, even if his is a bit bigger. The farmer makes a deal with somebody like Clint Houseman who has a big combine to do his threshing and probably come in and plant his corn. He has some livestock—Adam has beef cattle—and he and his wife both have part-time jobs, just as Adam has his ginseng business and Mary works in the delicatessen.

The second kind of farm is like Clint Houseman's, or Bud Petrescu's in Prairie. Enough land is rented or owned to make a $100,000 combine a paying investment, probably by doing custom work for neighbors, as Houseman does from time to time.

In both Prairie and Crow Creek, we found the value of farmland is about ten times the going rent. The range for rent is $28 to $40 in Prairie, $80 to $100 in Crow Creek. Property tax, as Adam complained, can be high. In Iowa it is $15 to $20 an acre, depending on the county. So if a landlord gets $90 per acre rent and pays $20 in taxes so his net per-acre income is $70, for the $1,000 it probably cost him to buy that acre, he is getting a 7 percent return. Most people do better with a certificate of deposit at a bank.

Without any doubt, American farming is facing its worst financial crisis since the Great Depression, and it will take a succession of good crop years to get out of it. In much of the country a farmer's debt-to-asset ratio is just under 25 percent. In 1985 farm debt was over $192 billion, compared with farm assets of $771 billion. In Prairie I found plenty of farmers with debt-to-asset ratios exceeding 40 percent.

Another big unknown, like biotechnology, is the price of energy. Experts—who predicted a third oil shock in the mid-1990s—since the Gulf crisis began, have started talking about a possible eventual "end of the oil era." Estimates of the world's proven reserves went up from 700 billion barrels in 1985 to nearly 900 billion by 1989. Roughly speaking, North America has 35 billion barrels, Latin America 122 billion, Western Europe 18 billion, the old Soviet bloc—mainly Russia—65 billion, Africa 56 billion, China 19 billion, and the rest of oil-poor Asia 22 billion. But the Middle East, led by Saudi Arabia and Iraq, has a whopping 571 billion barrels. In 2000–05, as the Western Hemisphere's reserve gets depleted, reliance on Middle Eastern oil and the Arabs who control it will go up dramatically, Saddam Hussein or no Saddam Hussein. So will prices. Pressure will be on to conserve, get better energy efficiency, and invest more in such alternatives as solar energy.

Robert Lounsberry, who retired in 1987 as Iowa's secretary of agriculture and was farming near Des Moines when I met him, predicts big farms will get into trouble with once again soaring oil-based fuel and fertilizer costs. Overly big farms, like corporations—there were just 7,140 in the 1980 census, fewer now—didn't fare well in the 1980s, most people said. An exception is Iowa State's Neil Harl, who argues the 1987 agricultural census shows corporate farms fared better than family farms.

The general feeling in Prairie and Crow Creek is that in the uncertain energy future the medium-sized family farm is most flexible; it can take reduced rations, hunker down, survive. A big corporation is far less able to adjust. If the red ink starts to flow or profits shrink, people say, the pressure is on to shift to something else. And what do you do with hired labor when it rains for 2 weeks? Who's going to work 16 hours a day to get the crop in if the weather's good? How many wives are going to go out and run tractors for free? Lounsberry says, "You can go all the way back to the Massachusetts Bay Colony. They tried to farm collectively and they about starved to death the first 3 years. So they divided up the land and let each guy that had a family keep what he raised. And they all seemed to grow a little more. So we have family farms."

There is one other way to keep people on farms, and that is to find new ways to use farm products. One proposal put forward by a group of professors at Texas A&M led by E. C. A. Runge, which takes into account the new oil shock and continued excess corn production, is to set up what they call a "sink" for surpluses, turning corn into ethanol. They figure Americans consume about 100 billions gallons of gas a year. To make that much gasahol—90 percent gasoline and 10 percent ethanol—would take 4 billion bushels of corn. Right now the United States produces a mere 800 million gallons of ethanol a year, so there is a long way to go. Professor Runge and his colleagues argue such a sink, flexibly tied to crop surpluses, would keep up American crop production, keep people on farms, and cut increasingly unstable oil imports.

Professor Harl at Iowa State told me that in the 1980s he reluctantly bulldozed down the barns and outbuildings on the farm where he was born. "We had 4 acres in which there was absolutely no economic usefulness. It was growing up in weeds and brush. The house itself went 10 years before that. My parents tore it down. Old homesteads are being demolished all over the country. If you tear buildings down in Iowa, you don't have to pay property tax on them."

And then, I said, all that is left is trees. As if waiting, forever, for people who will not be coming back. I saw so many abandoned windbreaks in Iowa and North Dakota. So much in the Middle West is going, going, and someday it will be gone.

"Yes," Harl agreed. He said he had about 2 miles of fences that needed cutting out. "Because trees have grown up around them. My dad let it happen before he died. It's a weed problem and the trees interfere with the big tractors and combines. That's the way it goes. Fewer and fewer people, less and less labor to produce food. My one tenant now farms over what was farmed by six families 75 years ago. It's almost universal. In China a local functionary near Shanghai told me his township was building a textile factory to absorb squeezed-out farm labor. 'We're gearing up to use our excess labor,' he said. Productivity goes up, so you use less land, less capital. You just don't need it. And that's what's been going on in this country from the beginning."

WHY, A READER in San Francisco or New York might ask, should I care if so many farms and small towns go under? What difference will it make to me? My answer is the city they live in, without the leaven of rural culture, is going to get a lot nastier—more family breakdown, more crime, more homelessness, more of a drop in America's emotional authenticity, a thinning of its kindness and good nature.

Most people feel this in a general way. Calvin Beale told me, "I believe the existence of intact small towns and farms has some importance to urban people as a nostalgic model of a simpler and more ordered society than the hectic, violent, polluted, traffic-ridden, polyglot metropolises in which they live."

Wayne Rasmussen warns, "The disappearance of farms and rural communities will, to my mind, reduce, at least to some extent, the traditional self-reliance combined with community support that characterized rural America. There will be an even greater emphasis upon 'I' rather than upon 'we.' All virtue does not reside in the countryside nor is all the countryside virtuous, but I believe that there was a hard kernel of truth in Jefferson's agrarianism."

Gilbert Fite, also looking at it as a historian, predicts our transition from a rural to urban people will fundamentally change American life and character. "Precisely the kind of society that will emerge when rural influences are finally gone, except in myth and history, is not yet clear," he told me. "But I believe the new society will be different from that nurtured in the openness and freedom of the farm and small town. It will be, I think, a society with less social and neighborly concern, one much more suspicious and less trusting, less honest, more greedy and more concerned with self-indulgence than service, and less concerned with or governed by any basic moral principles."

He, too, mentioned Jefferson. "There cannot help but be a difference in a society of people raised on farms and in small towns and those raised in the suburbs and poverty-ridden inner cities. I believe there is something to the Jeffersonian concept that living close to the land and in a rural environment has an effect on people's thought, character, and actions."

Norman Borlaug, as a scientist, sounds a similar warning. When it comes to the computer age, he feels, "There is a tendency around the world to substitute these mechanical gadgets for the human brain and creativeness. And I think it's a step back toward worshipping the golden calf. I sometimes worry about American civilization. This revolution in social values that is going on. If it defeats our will to work, if our work ethic goes, what sort of society will we have?"

Can we be more specific about these rural cultural influences, why Jefferson called farmers "the chosen people of God"? I think we can. In 12 years of systematically looking at villages in Asia, Africa, and Latin America, and in 1990 spending some months in a village in Poland, I found certain universal traits. Many of them I listed early on, when describing nineteenth-century rural American communities. Rural America today is radically different, of course, in its oil-fueled machinery and comparatively empty countryside. American rural people are also more subject to urban culture in the form of television, movies, computer-age information technology, education, and travel.

But when it comes to the rural cultural traits that really count, such as those to do with family and property, I find them almost as strong among the farmers of Prairie and Crow Creek as I do among village people the world over.

Everybody who farms, affected as they are by too much or too little rain, or wind, hail, frost, snow, or whatever, has to be fairly fatalistic. The agricultural setting reinforces religion. Churchgoing—and I mentioned 77 percent of rural North Dakotans go to church on Sunday—with its rituals for birth, marriage, and death, and such holy days as Christmas and Easter, confers meaning and dignity on rural lives. Sunday school, when it comes to culture, matters a lot; its stories of Moses in the bulrushes, Joseph and his coat of many colors, Samson and Delilah, David and Goliath, the Good Samaritan, both teach morality and enrich our literary sense; what you learn by the age of 12 stays with you forever. When it comes to religion, the fatalism I mentioned, combined with skepticism toward organized religion (jokes about the pastor) and deep personal faith, are as common to a North Dakota or Iowa farmer as to a Punjabi or Polish peasant. Rural people are plain, straight, and conservative in outlook the world over.

The farm family in Prairie and Crow Creek looks to the father to provide food, shelter, and clothing; his authority has an economic basis. Each of its members in turn is obligated to work for the family under the father's direction. In Prairie we find this to be true of the old-fashioned Zuckermans, who carry traditionalism to the point of eccentricity. But it is equally true of such successful modern farmers as the Petersons, Fischers, Zimmermans, and Hansons. Two exceptions are Crow Creek's Clint Houseman and Prairie's Bud Petrescu, the Vietnam veteran. Both have been subject to outside, urban cultural influences. Both farm extremely big, wholly rented holdings and work very hard, using big machines. Both have the help of their wives, liberated young women who drive a combine or a tractor with all the skill and endurance of a hired man.

Adam Pierce and Miles West in Crow Creek are perfect examples of the authoritative father (though Mary Pierce feels Betty West is at times too submissive). Their families are of central importance to both the Pierces and Wests; blood ties and kinship have heavy weight. The us-and-them attitudes of Pierce and West toward the urban commuters in Crow Creek, and the kind of exaggerated exclusivity we find among the Zuckermans in Prairie, to me are examples of people carrying the institution of family too far. But Pierce, West, and Zuckerman all feel culturally threatened and, like pioneers fearing an Indian attack, circle their wagons. (Remember Mrs. Schumacher saying that if the Zuckerman girl and boy don't get off the farm, they won't meet anybody to marry.)

When it comes to marriage, Americans, rural or urban, are much more romantic than one finds in the Third World. In Chinese or Indian villages the approach to marriage is practical, even carnal, whereas among Americans the idea of romantic love is practically universal. Over the years many American farm marriages evolve into a kind of dogged partnership of closely related work life and family life. Deep affection can grow out of compassion and mutual need. But a peculiarly American cultural trait is our high standard of conscious happiness. Marriages can fail when people feel they are not specifically "happy," a discovery that has broken up a good many American homes.

While the monogamous, divorceless, multichild marriage is still held up as the ideal in Prairie and Crow Creek, it is being challenged by economic reality and urban America's sexual revolution. From 1960 to 1975 this country saw a huge rise in postpill premarital sex. In the same 15 years the birthrate fell by 38 percent and the divorce rate doubled to half of all first marriages, 60 percent of all second marriages. Full-time American home-makers dropped from 75 percent to 25 percent of all married women. In 1971 half of all young American women were married by 21, men by 23. By 1990 the age of marriage of both men and women was up two-and-a-half years.

The impact of all this on the American family has been devastating. One-third of children born in the 1980s may live in a step-family before they are 18. One out of every four is being raised by a single parent. A fourth of today's children are born out of wedlock, a third to teenage mothers. Poverty claims 20 percent of all children, 40 percent of black and Hispanic children.

All these trends are, of course, much stronger in urban America, if present in rural communities too. Yet in both Prairie and Crow Creek there are still plenty of farm families who fit the old Norman Rockwell image of Dad plowing in the fields, Mom canning in the kitchen, and tiny tots or school kids helping out with the chores. In America at large, fewer than one family in ten now fits this traditional pattern.

Marital fidelity is also stronger in rural America, partly because there is so much surveillance in a small community. In a city, affairs can be hidden in the protective anonymity of the crowd. Rev. Andrew M. Greeley, the Chicago priest and novelist, argued in a February 1990 report that fewer than one in ten currently married people have been unfaithful to their spouses. This contradicts the findings of Dr. Alfred C. Kinsey, the famed sex researcher, who found in the 1950s that half of the married men of the time and a quarter of married women had had extramarital affairs. But even Father Greeley's 10 percent figure suggests marital infidelity is wide-spread.

More than 40 percent of Americans now live alone. This partly reflects the freedom of choice made possible by wealth and technology. In Third

World villages practically nobody lives alone, one person being such an unviable economic unit. (Who gathers wood, fetches water, tends the fire, and cooks the meal while you are out tilling the fields?) The only ones I met in Prairie who lived alone are the Catholic priest, Father McNeeley, and a few elderly widows. In Crow Creek there is just the recently widowed Louise Fletcher who lives alone, though Alice Mitchell thought of another elderly woman and remembers in the 1940s Crow Creek had a lot of "old ladies living alone."

Both Prairie and Crow Creek have broken with the near-universal rural tradition of welcoming more children as more hands for work and more security in old age. As farms mechanized in the 1930s and 1940s, this cultural trait lost its economic basis. Farm couples no longer expect to have bigger families than city people. Calvin Beale says a national survey by the Bureau of the Census in 1987 showed that farm women 18 to 34 expected an average of 2.3 children; for urban women it was just 2. But as recently as 1960 the farmer's wife would have expected 3.3 children to the urban woman's 2.3. In 1960 farm families averaged 85 children under 15 years old for every 100 adults of prime work force age, partly reflecting the baby boom. This fell to 44/100 by 1987.

Today's life expectancy of 75, up from 47 in 1900, as mentioned, is mainly due to falls in infant, childhood, and maternity mortality. The trick in 1900, especially on a farm, was to survive into adolescence and, for a woman, not to die in childbirth. In much of the Third World it still is. And in spite of our ability to cure more diseases, replace more organs, and prolong life, the average 65-year-old American male in 1991 can expect to live just 3 years longer—his wife, 6 years—than he could at the turn of the century.

We often hear that one reason for the Third World's population explosion is the old desire to have more sons for field hands and as security in old age. (The biggest reason in reality is the lack of schooling and better jobs for women.) Actually, in most villages, there is a distinct put-down-tools attitude among much younger village fathers once their sons are grown, able-bodied, and not in school.

We're not so different. Most American farmers would like to see a son take over their farm, though it is later, usually after 4 years at a university or 2 years at a technical college. Overall in America in 1990, just 75 percent of American men aged 55 to 59 were still working. Only three Americans work for every retiree. My guess is these figures hold true in both urban and rural areas.

The experts say the biggest number of future farmers will hold part-time jobs. But off-farm work itself is changing. Full-time jobs, the skilled kind that pay $10, $14 an hour, are harder to come by. There are more $4- or $5-an-hour part-time service jobs. Or temporary jobs—in Prairie young men go out on crews with a local outfit that salvages trains all over the north central states; since there is a lot of overtime at $9 an hour, the pay is good, but the work doesn't last long.

Part-time jobs are already common. Charlie Pritchett's son Jack, you'll recall, works as a crop appraiser, a taxidermist, and a seasonal helper at the fertilizer depot and raises sheep, ducks, geese, and a few Arabian horses. John Farrow, besides farming, sells haying equipment and is a seed company's area supervisor. Emma Muhlenthaler's nephew farms and works as the elementary school's part-time principal; he used to deal blackjack at the Cabaret out on the highway but decided it set a bad example. Part-time work is very common in Third World villages too, among men and women both.

Mary Pierce's part-time job in a city delicatessen, part of a chain supermarket, is typical of the growing number of service jobs that pay $4 or so an hour, just over the minimum wage, and avoid paying retirement and other benefits by limiting the hours. Women I interviewed had mixed emotions about such jobs, as they are both exploitative and convenient. Mary, for example, works just 28 hours a week; she gets off at 2 p.m., which gives her time to shop and do her housework and prepare the evening meal for Adam and the two children at home. I talked to her about it, and Mary wants to keep up the breadwinner-homemaker family tradition as much as she can; she even makes Adam's lunch, leaving it in the refrigerator. Women like Mary now make up 56 percent of the American work force; they average just 70 cents for every dollar made by a man.

With pay so low and so many part-time service jobs available, a growing number of women have two or more jobs; their numbers quintupled, from 636,000 in 1970 to 3.1 million in May 1989, according to the Bureau of Labor Statistics. This revolution in part-time service jobs mainly affects women; in 1970–89 moonlighting men just went from 3.4 million to 4.1 million. No urban-rural breakdown was given, though the proportion seems likelier to be higher in cities where more jobs would be available.

Village women work hard the world over. They do 60 percent of the farming in Africa and more than half of it in Korea and China. And I've never seen women do such hard field labor as they do in Poland. Since women invented farming, as most historians assume, this is not so surprising. In Kenya I've been in villages where men still think of themselves as primarily hunters and herders; growing food crops is women's work.

Many American women, and men too, find all the new part-time service jobs, and the need to take them, one of those unpatterned, unsettling, and confusing changes brought by advancing technology. Everybody is coming to recognize that what counts today is not muscle power but skill. In both Prairie and Crow Creek we find very high rates of college attendance, either at a 2-year technical college or a 4-year state university; this is true of almost all the young men going into farming. Others are aware that if they go to a university and do not take agronomy or some other farm-related degree, but instead learn a profession, they are in effect educating themselves to leave home and go to a city to live. This is often done reluctantly by youths who, if there were any opportunity, say they would go into farming. Others, like Tom Pierce in Crow Creek and Kurt Schumacher in Prairie, deliberately get a 2-year technical diploma in some skill like audio equipment repair or auto mechanics so they can get a job close to home. Both hope to farm part-time.

The way so many young people in Prairie and Crow Creek are going to college pretty much does away with the old idea that country people are hicks and hayseeds. Interestingly, it is the second- and third-generation Norwegians and Germans in Prairie who speak standard American English, if sometimes with a trace of Scandinavian or German accent. Crow Creek's natives, mostly of English descent, can sound more folksy, with the occasional *ain't* and *come* for *came* and such phrases as *in a pickle* or *reckon,* in the sense of *guess* or *suppose.* More Old English, biblical, and Shakespearean phrases have found their way into the Iowans' language, even among university graduates, giving it a rustic sound.

In the farm families of Prairie and Crow Creek one finds the universal rural belief that a strong work ethic has to be inculcated from early childhood; children become self-reliant by performing useful chores as toddlers. On most farms we see children eager to perform farm work, especially if a skill or machinery is involved, as proof they are growing up. Bud Petrescu's 12-year-old son is as proud to drive his father's tractor as Tom Pierce is to show Adam he can fix the combine. Boys and girls are eager to work side by side with their parents; one finds whole families harvesting.

The strong desire of children to follow their parents' footsteps, a universal trait, is often frustrated in America by so few opportunities to farm. Like Prairie's Chuck White, they are forced to learn another occupation.

Calvin Beale says a national survey taken in 1973 showed that 81 percent of male farmers 21 to 64 years old were sons of farmers. He feels this may have gone down some today, but not much. No nonrural occupation remotely approaches this father-son tie that we find among farmers.

Of course there are plenty of exceptions in our urbanizing society. Robert Williams said his sons didn't like farming, milking, or chores. Some farm kids get so burned out from overwork they never want to see another farm. Judy Pierce vowed she would never marry a farmer because she saw how hard her mother worked without pay. This flat rejection of old values naturally can cause some generational conflict. In Judy's case it is leavened by her two sisters' marriages to a dentist and a banker. Even if the second farms part-time, these sons-in-law have brought more urban attitudes into the family.

A change in favor of farming has been the rise in comparative living standards. Janet Schumacher recalls when she graduated from high school in 1958, all her girlfriends wanted to get off the farm because in those days there were so few amenities. "You know, they *hated* the farm. Then. It wasn't modern enough. You never had running water. It's true there were some comforts by then. We did have electricity. Several had indoor plumbing. But my family didn't. We did get it in 1960. And, you see, a lot of them milked cows yet. And everybody hated that. You had a few chickens and you had to milk a few cows. And the kids all wanted to get away from that. Go to town. Where there was green grass." She laughs. Of course she chose to stay back on the farm herself.

The Schumacher house is big, rambling, and comfortable; with its lawn and flower gardens and modern ranch-style architecture, it looks like what you would find in any prosperous urban suburb. When I went there for dinner, they served wine. "So many of these young couples now," Mrs. Schumacher says, "they want to buy a big old-fashioned country house. Back in my day we couldn't wait to get rid of it."

In spite of the comforts, whether modern commercial American farmer or subsistence Third World peasant, hard physical work, as Clint Houseman reminded us, is the central fact of life for anybody who makes his living from the land. Attitudes Robert Redfield found among Mexican peasants are just as true of the Prairie–Crow Creek farmers: the feeling that farm work is good and commerce not so good, that skill at cultivating and growing and maturing crops, and running and repairing machinery, reflects maturity and a sense of personal worth (Peterson, Fischer, and Schumacher are held up as "good farmers"). While everybody is glad when the hard work is over, whether in the Punjab or Middle West, there is a marked rise in morale during times of intense field work, such as a harvest. There is corresponding disorientation and demoralization during periods of prolonged idleness.

In both communities you also find the feeling, common to most Third World villages, that their moral code is somehow superior to that of the cities (Carl Sandburg's "painted women under the gas lamps luring the farm boys"). The thrifty, hardworking neighbor is praised (Fisher, Schumacher, Houseman), though there can be criticism if it is felt a farmer has gotten too big and that his prosperity is gained at the expense of others. George Foster tells me the old sense of "limited good" is fading away as villagers modernize. But you still see it in villages in England.

In Prairie one hears criticism of Ernest Waldenberg. How, people ask, can anybody have over 9,000 acres, farmed partly by hired workers, without it being at someone's expense? Resentment was voiced when word went around that Waldenberg got $300,000 in debts written off. Generally, however, there is less envy in rural America than anywhere I've been. There is so much opportunity in our country, the economy is seen as a deep well to be forever primed and pumped. And not as a fixed pie with just so many pieces to fight over—the feeling in many societies.

There is a distinct Old Testament quality to the agricultural moral code, in both villages and rural America. Rural restraints, conventions, and patterns of obedience to authority are much more in tune with the stern Ten Commandments than the loving Sermon on the Mount. Indeed, in the main civic square of Fargo, there is actually a stone monument engraved with the text of the Ten Commandments, a gift from the local American Legion post. Such strictures as "thou shalt not commit adultery" and "thou shalt not covet thy neighbor's house, thy neighbor's wife . . ." look a bit startling in this day and age.

RELIGION, I KEEP stressing, is the core of any culture. There is no doubt that as the West grows richer and its science advances, the spiritual dimension of life becomes less. Yet even those Americans who do not profess to be Christians or Jews, or, for the few, Muslims, Hindus, Confucians, or whatever, are shaped by ethical standards set down centuries ago by the great religious prophets and teachers.

Again, as I have said, when it comes to practical day-to-day living, this inherited ethical system and the coherent meaning it gives our lives affects everything we do: our industry, inventiveness, initiative, enterprise. In Britain when Margaret Thatcher exhorts people to forsake "a dependency culture" in favor of "an enterprise culture," in the next breath she tells them to go back to church, to go back to "Victorian values," which, before two world wars and loss of empire, were probably the last coherent value system the British people had. Her religious appeals make some of her political supporters squeamish, but anthropologically speaking, I think she is sensible to go right to the heart of the issue.

Here I'd like to turn for a moment to a few older Americans, who, because of their race and rural origin and experience in life, seem to me to have an especially strong core culture. By that I mean a metaphysical belief system more in tune with the universal religious and moral norms found in the world's villages than many of the rest of us do. And their culture, much more that the culture of Prairie and Crow Creek, is under attack from what I regard as the most anticultural urban influences: secular and market-minded norms, with expedient relations between buyer and seller and priority given to making money. What they tell us, I think, is that religion and its supernatural supports are necessary to morality—that is, a modern, urban, commercial ethic may be too weak to withstand the savagery that lurks below our civilization and emerges in our dreams, crimes, wars, and, these days increasingly out in the open, in our horror films and television.

When I compare religion in America with the deep personal faith often found in villages, the nearest thing I find to it here is not among rural whites but among blacks of rural Southern origin or Native Americans. By this I mean a deep belief in a personal god concerned with one's own individual welfare.

In Waterloo, Iowa, an industrial city of 75,000 a couple of hours' drive from Crow Creek and home of John Deere's biggest tractor plant, Martha Nash, a gray-haired woman who runs its Martin Luther King, Jr. Center, described what she feels is the distinctive quality about much black religion. "Blacks brought their abstraction of God from Africa, and I think it stayed with them. God in African tribes is not anthropomorphic in the Jewish way. They kept that very personal relationship with God." Mrs. Nash says the church fills a social need for black Americans as a place where they can express their talents and leadership, pretty much away from white society. "In a church you can control. That's the outlet for them. That's very important to them. We haven't had much of your nice little happy family." Ruth Anderson, a professor at Northern Iowa University in Waterloo, told me, "For a long time the black church was the only way Western society let us in. Everything we did, we did within the church. Religion stayed personal. As black people we say, 'I'm going to Heaven for myself. The preacher can't go to Heaven for me.'"

Ada Tredwell, a coworker of Mrs. Nash whose father was a poor sharecropper in the South before coming to Iowa in 1916, put the breakdown of culture, as a set of rules for living, into a few words. "In hard times more young people come to church. They're worried and confused. The way values keep changing, they don't know what the rules are. There was a time when, if a boy got a girl pregnant, he had to marry her. That was the way we were brought up. Now you look at television and say, 'Hey, that's okay. This is okay.' And you find yourself changing your life, your standards, and your values.

"It keeps getting foisted on you. You see it enough times, you keep seeing it over and over. And you think: what's wrong with that? So we accept it. There was a time you protested, 'Oh, I think it's terrible these kids are getting pregnant or these babies are having babies.' Now people criticize you for even objecting. And naturally the more you accept it, the more acceptable it becomes. And you lose your values. That's what happens."

More than any other group in our society, blacks have gone from rural to urban the fastest. In the early 1940s, three-quarters of all American blacks lived in the South, most of them rural. Then, as mentioned, 95 percent of all black-operated tenant farms—about 350,000—went under between 1950 and 1970. (White tenant farms dropped by 70 percent; about 1 million.) The already small number of farming blacks were halved again in the 1980s, from 243,000 to 123,000, as older farmers went broke, retired, or died. Where did the agriculturally dispossessed blacks go? In 1940–80 Chicago went from 8 percent black to 40 percent, Detroit from 9 percent to 63 percent, Washington from 28 percent to 70 percent.

The cultural consequences of this very sudden rural-to-urban shift have been devastating. Blacks are now over twice as likely as whites to be jobless. The median black family income is 56 percent of a white family's. Nearly a third of blacks, as against 10 percent of whites, live below the official poverty level; among them 45 percent of all black children, as against 15 percent of white ones. A newborn black baby is twice as likely as a white one to die before its first birthday.

This country's 31 million blacks, 12 percent of the population, learn less, earn less, live worse, and die sooner than the rest of Americans. A black man is six times as likely as a white man to be murdered; homicide is now the leading cause of death for young black men and young black women both. One young black man in four, penal researchers say, is either in prison, on parole, or on probation.

The sudden black American rural-urban shift might not explain everything. But when you uproot an essentially rural farming people from the South (and their African heritage is rural too) and expose them to the ugliest, loneliest, most crime-ridden and poorly educated slums and possibly being brought up by a teenage mother on welfare in a crowded apartment, and repeat that for a couple of generations, it is a wonder so much of their culture survives. It is not a question of race; in Britain, where you get the same cultural breakdown among uprooted rural people trapped in city slums, the victims are practically all white. Indeed, plenty of whites in America fit the same pattern. But whites do not suffer as much social isolation as blacks; for some blacks, their main contact with the white society around them comes from television.

It is just common sense that the more violence you get in movies and television, the more violent your society is going to get. As T. S. Eliot said, "An artist has influence over us whether he wants one or not and we are influenced whether we want to be or not." While I was in Britain, a young man in a Rambo-style bandana and bandolier shot up a shopping center, killing people. Mrs. Nash: "If you keep drumming it into people in their home, you're telling these kids, 'This is the way you solve problems. You don't talk it over. You don't reason. You just blow 'em away.'" Mrs. Nash warns, "America is in for a lot of social upheaval, social disintegration."

Few white Americans, as yet, feel so culturally threatened. In Prairie I found the one person who did was John Running Horse, the earlier-quoted 80-year-old Sioux Indian. He also has a deeply personal religion. "Sometimes I go out in the woods and sit down and smoke my pipe," he told me. "It's stone and a piece of wood. The feather at the end of the pipe, that's a symbol of long life. We use it only for meditation. All people ought to have the right to worship the way they want to. But the white man says, 'That's wrong!' That's all right if he wants to think so. But in time he can cause an Indian to believe it's wrong."

Around Hawk County, Sioux seem to get caught up in a vicious downward spiral: the worse whites think of them and the lower their self-esteem, the worse they behave. The otherwise often admirable Sheriff

Jager exemplifies fairly common local white prejudice: "The boss of the carnival out at the county fair this year, he had a lot of Sioux. Now if a boss knows how to control his people, he'll pay 'em $10 cash to eat, sleep, and live on. So they don't have nothing to get drunk on. Then at the end of the season they get a big payoff. But this guy, when some of 'em got mad and quit, he gave 'em money. And they put on a heck of a drunk.

"I had two or three of 'em we had to haul 'em out of town, get rid of 'em. Dump 'em off in other counties. They'd just get drunk, fighting, raising Cain, lying around Main Street, passing out. The Sioux, they're the meanest Indians. If you ever wanna get cut. If I got a Sioux and a Chippewa and if I wanna get cut, it's gonna be the Sioux that's gonna cut me. They're a violent Indian and they're the drunkenest bunch you ever saw. Those other tribes, they're not as crazy as this bunch you've got here. If you look at the murders on the Indian reservations in the state, a majority of 'em are on the Sioux reservations, like up in the Fort Totten–Devil's Lake area. The violent crimes where they'll stab one guy or run over him, eight, nine guys to kill one guy, just overkill."

No wonder—and such attitudes are fairly common—the pressure can get too much. Running Horse showed me his grandson's picture. The boy was the jockey who got thrown at the county fair. "He was staying here with me," Running Horse said. "Putting his salary into a savings account. He was all set to go down to Concordia College. But he couldn't hold up under pressure. He started drinking. He'd work until eight at night at the filling station, and from eight he was out. So I talked to him and I said as long as you're involved with this kind of life, I'm not doing you any good. So he found another place. Where is he now? He's uptown, I guess. He was going to college too. If he keeps on, I don't see how he can go."

When we look at outside, urban influences on rural American communities, I feel drugs and exploitative horror shows, which make the unthinkable thinkable, are probably the worst. Sheriff Jager and Miles West talked about drugs in Prairie and Crow Creek. One wonders what to make of the problem, reaching down, as Jager said, into the grade schools. America seems overwhelmed by drugs. The Institute for Social Research at the University of Michigan, which every year questions as many as 17,000 of the country's 2.7 million high school seniors, both urban and rural, found 50.9 percent in 1989 said they had at least tried an illicit drug like marijuana or cocaine, as against 53.9 percent in 1988 and 56.6 percent in

1987. Just 3.1 percent said they used crack in 1989, the same percentage as in 1988.

While the overall trend is slightly encouraging, critics say the figures probably understate the problem, as they do not include high school dropouts—now running at more than 27 percent. Dropouts, like all unemployed, have the highest narcotic use rates. When it came to cigarettes, smoking was down to 29 percent of the seniors questioned, a rate that held throughout the 1980s. Drinking was down; 60 percent of the seniors said they had consumed alcohol in the past 30 days, down from a peak of 72 percent in 1980.

As I've said, I'm baffled by the drug problem. Little farming towns seem no more immune than the most poverty-ridden inner city. In Third World villages a few drugs appear; but a rampant drug problem like ours I've only encountered abroad in a few Western cities like Liverpool or Amsterdam.

When it comes to the commercial and exploitative use of horror, let me quote a story I heard from Sheriff Jager on my last visit to Prairie: "One night the Harvey Police Department called me and they said, 'We've got a crazy.' On the radio I can hear what sounds like a dog growling in the background. Well, here it's this lady they had. I mean, she literally growled and howled. She was a Devil worshipper. She had the upside-down crosses, all that stuff. So I drove up to Harvey hospital. No reflection on ministers, but there sitting in the hospital entry is this minister from one of the smaller new churches. And he's got his hand on her head. And she's on all fours like a dog, howlin' and caterwaulin.' And he's hollerin', 'Cast out the Devil! Cast out the Devil!'

"Just like on TV, you know. You see those things and you don't believe it. And she's snarlin' and making all sorts of weird sounds. So I go and say, 'Lady, come on, enough of this bullshit. You gotta go home or I'm gonna put you in a padded cell. You're crazy.' It turned out she was in a Satanic cult in Minot. She had a man's voice and said she was possessed and was just as goofy."

Sounds like *The Exorcist,* I said. ABC News reported in March 1990 that polls show two-thirds of Americans say the Devil is real, half of them that he has influence on their lives. ABC's Jeff Greenfield was taken aback. "Maybe," he said, "the Devil provides an explanation for things we don't understand." ABC said pop Satanism, rock singers like Ozzy Osborne, and Satanic youth cults were on the rise.

"Yeah, just like *The Exorcist*," Jager told me that day. "We're finding more and more of it coming back in. There's a great big cult out of Minot.

Devil worshippers or whatever you want to call them. A lot of it hooked through the airbase. Matter of fact, we had a double suicide over in Sheridan County 2 weeks ago. Two boys shot themselves. Deer rifles. And it turns out they're affiliated with the cult and black magic."

OVER THE YEARS, coming home on visits from the Third World, I used to feel in a vague, formless way that the people to worry about were not them, but us. In the villages of Asia, Africa, and Latin America, man's basic institutions of family and property are still strong, as is religious faith and the agricultural moral code.

This is no longer true in the urban, postindustrial West. I found, while writing my Britain book, that so much of its predicament is the same as ours: the same shrinking manufacturing base, failing schools, family breakdown, social polarization, homelessness. In her 1990 book, *May You Be the Mother of a Hundred Sons,* Elisabeth Bumiller, a former lead writer of *The Washington Post*'s Style section, writes that she expected, when she began what became a 4-year stay in India, that what she learned would have "little consequence for the lives in the world from which I had come."

She goes on, "But slowly I realized that the way Indian women live is the way the majority of the women in the world spend their lives; it is Americans who are peculiar. Ultimately, I realized my journey to India was a privilege. Rather than going to the periphery, I had come to the center." I have recently felt like this in a Polish village—one is back in the world of reassuring universal values.

In the 1980s whatever is going wrong in America began to take more definable shape. But it is not just happening in America alone—rather, I suspect, in all societies on the cutting edge of Western society's forward technological thrust. When we look for local, usually economic, explanations, I think we are making a mistake. For example, why did so many people begin living in the streets in the 1980s? We can say that misguided reformers threw tens of thousands of patients out of mental hospitals under the "deinstitutionalization" policy. Or that we fail to provide adequate mental health care. Or is it because the law prevents police from taking derelicts off the streets? Or because a shift in the real estate market eliminated millions of cheap rooms? All plausible, but none get to the heart of it.

For homelessness, I believe, is just one symptom of a profound change caused by the diminishing rural base of our urban culture. Moreover, it begins to look like a wider Western phenomenon. Certainly, great changes in the general human condition are happening in both America and Britain at the same time.

Economically, both societies are polarizing. People engaged in growing or making things, like farmers and factory workers, are getting poorer. The beneficiaries of the new high-tech, finance, scientific, and information industries, plus the political and cultural elites, are getting richer. Greed-is-good ethics make fortunes as fast in the City of London as they do on Wall Street.

In America we also see rural-urban polarity; in 1985 the poverty rate for rural America was 18 percent, in cities just 12.7 percent. Overall, average real wages in this country have declined since 1973, since 1985 alone by 5 percent. A January 1990 New York Times–CBS poll showed 36 percent of its respondents (43 percent of women alone) felt the economy was getting worse. We have the same failures in schools, in health care, in finding practical and useful applications of our scientific discoveries.

But homelessness is a direct result of the breakdown of urban family life. America has about 4 million homeless in all, Britain a quarter that many. They include the mentally ill, alcoholics, drug addicts, destitute old people, drifters, and just helpless people who fall through the supposed safety net. A family could have rescued them.

Between 500,000 and 12 million of them in this country are "runaway" or "throwaway" teenagers. Preliminary results of a federal government study in February 1990 showed the homelessness of most youths is caused by family breakdown. Nearly 60 percent report being physically or sexually abused by parents or other family members.

In inner cities on both sides of the Atlantic you also have an emerging underclass, heavily black in America, mainly white in Britain. This revives Oscar Lewis's old theory that if anybody stays poor and jobless long enough, they sink into a "culture of poverty." Lewis defined this as a "defense mechanism without which the poor could hardly carry on." He said it had seventy identifiable traits. One was "frequent use of physical violence in the training of children and wife beating."

We talked about the universality of rural culture and looked at what Prairie and Crow Creek had in common with Third World villages. Robert Redfield, the University of Chicago's great anthropologist, once compared three rural societies: the Mayan Indians of Yucatan, nineteenth-century English villagers, and ancient Greek cultivators. He found life and culture so alike, he speculated that if a villager from any one could have been transported to any one of the others and somehow knew the language, "he would very quickly have come to feel at home." This would be, Redfield said, "because the fundamental orientations of life would be unchanged." But imagine dropping that villager among the homeless, mentally ill, or trapped underclass of an American city, with all the hunger, cold, violence, and loneliness to be faced.

The University of Chicago's Theodore W. Schultz wrote me just as I was completing these final pages. "Fewer farms and fewer towns—what does it mean?" he asked, repeating my question. "Neither Adam Smith nor Karl Marx had answers. Long books notwithstanding."

Professor Schultz, who was born in 1902, has personally seen most of the big changes in rural America. "In 1910 the farm population was 32,077,000, glory be," he said. "In 1987 the count was down to 4,986,000. Even so, half of farm family income is now earned doing nonfarm work." Having watched rural America fall from 35 percent to something well under 2 percent, Professor Schultz feels wisdom and the long view are no longer "compatible."

All culture, as I said from the start, has an economic basis. If we want to preserve the rural roots of American culture, the place to begin is with agriculture. The problem boils down to how to keep sufficient numbers of people rural—and in farming. As we see in Crow Creek, it is not enough for people to simply live in a rural setting, as in a rural bedroom community. They lack the evidently essential organic tie to the land that a farmer has.

So we go back to economics and trying to solve the farm problem in terms of production, prices, costs, net income, and so on. And more than just economics: politics, climate, foreign policy, and what happens in Russia, China, and India, the environment, oil, eating, drinking, and smoking habits, all kinds of things that figure today into what happens on an American farm. And what is good for one farmer is maybe not good for another. High grain prices are not good for chicken growers, or hog and cattle feeders.

A range of opinions from respected experts has been quoted on ways to deal with the farm problem. They all agreed that the point and purpose of American farm policy is to keep producing plenty of food for ourselves and export at prices fair to large and small farmers alike.

Everybody agrees it would be a good thing if farmers were less dependent on crop supports. As in the 1990 farm bill, there is a gradual trend to lower price supports, less government intervention, and more flexibility in farming. Under the subsidy program, which applies to fourteen farm products, the government still pays cash to farmers for every bushel they produce—the payment of corn, for example, is about $1 a bushel—in exchange for a specific amount of land being left fallow.

Environmentalists say, as does author Richard Rhodes, that trying to get as high a yield as possible on what land remains encourages heavy use of chemicals and can lead to severe soil erosion. They seek to limit the use of chemical fertilizer and pesticides as a way to protect water supplies, fresh food, and wildlife.

For the next 5 years, farmers will be getting incentives to farm in more environmentally friendly ways. The earlier quoted Kenneth Cook of Washington's Center for Resource Economics says organizations like his own came to realize taxes to discourage the overuse of pesticides, herbicides, and other harzardous chemicals would not get through Congress in the 1990 election year. To get what they wanted, Cook says, environmental lobbyists had to face the fact that "agricultural committees are not big on telling farmers what to do, but are big on giving them money."

So they proposed measures to protect the environment by giving farmers entitlements to new government subsidies: entitlements that had the added advantage of being popular with the public. This tactic worked. The Conservation Reserve Program in the 5-year farm bill, which takes land out of production, was expanded. Farmers are given incentive payments to prevent chemicals from seeping into the groundwater.

Environmentalists have even gone along with easing penalties on farmers who drain and plant protected bogs and marshes. Cook says the change might actually "enhance enforcement." Under the old law, farmers could lose their entitlement to all farm benefits if they grew on wetlands. But few offenders were prosecuted; local agricultural officials proved unwilling to ruin farmers financially for filling in swamps and ponds. The environmentalists are expected to make an even bigger push in 1995, again relying on incentives rather than prohibitions.

President Bush is under increasing international pressure to join in coordinated action of the global environment, with a rift between America and the European countries on the need for action to mitigate global warming. Bush is also under Third World pressure to end all farm subsidies deemed to distort trade, something the Europeans refuse to do.

Everybody agrees, at a time of deficits and budget stringency, that somehow the cost of the farm program has got to be brought down. Since America is a democracy, it pretty much rules out radical solutions. We can't impose ceilings on land or on the size of mechanical equipment, even if either might save most family farms and small towns. To be realistic and pragmatic, we will almost certainly keep target prices for corn, wheat, rice, and soybeans, and some modest supports for cotton and milk. The bargaining power of farmers is too weak, in our free-market economy, to do wholly without price supports (much as farmers like Adam Pierce or Robert Williams would like to). The trade-off for price supports is acreage restriction to avoid large price-lowering surpluses.

"There is no use talking about some ideal solution to the farm problem," says Gilbert Fite, voicing what seems to be the consensus. "We have had 60 years of experience and history of farm policies from which we cannot escape. Even if we had a clean slate, that is, no historical experience to affect our thinking, I doubt if we could develop an overall farm policy that would be much better than what we have.

"But we don't have a clean slate. We have a history of conflicts, habits, and special interests that have grown into the system. So, in conclusion, realistically, I would tinker with, and modify, the present farm policies and try to bring about improvements where possible in the real world in which we operate. This isn't much of a solution, but it's the way democracies work."

The radical change I would like to see is not so much in farm policy, though I would modify it as much as possible in favor of the small farmer, but rather in the way urban people perceive the rural crisis. This will not be easy because there is no immediately obvious cause and effect. The effect of loss of farm life is over generations and so is the gradual erosion of such rural-based cultural institutions as family and property. But, as I have tried to show, if we get ever-bigger and fewer farms, and thousands of small towns die, and with them their churches, schools, and businesses, the greatest cultural loss will not be to the dwindling minority left on farms— their culture ought to stay pretty much intact—but to the growing urban majority of us in the cities. As Professor Fite reminded me, "This is the last generation of Americans, we older people, who will have any significant hands-on experience with farming, land for farming, or a real agricultural life. Those left are so few they will no longer be numerous enough to provide any farm leaven in our society."

Cultural momentum continues long after a culture loses its economic basis. The ethics of an Ada Tredwell, for instance, nurtured in the rural

South by her sharecropping, churchgoing parents, are still a leaven among some lives she touches in Waterloo's inner city. But not indefinitely. When she is gone, some of this momentum will be lost. As Richard Leakey says, we are each innately capable of being anything, from a Nazi to a saint. As cultural animals, we are the product of our culture, even if it's a culture of poverty. Given enough time in some desperately poor slum, without any counterinfluence from family, church, or school, you will find dropping out, drugs, crime, dependency, teenage out-of-wedlock childbearing, and family breakdown becoming a kind of behavioral norm.

As I said at the outset, the world's villagers and ourselves all exist in the same continuum stretching unbroken through time. Elisabeth Bumiller describes an Indian village: "The ways of the 1,000 people of Khajuron are the ways of most of humanity." We stand at the farthest end of this continuum, the most technologically advanced society ever. But what we find most culturally meaningful is deeply rooted in the distant past. Remember F. Scott Fitzgerald's famous ending to *The Great Gatsby,* when he says Gatsby believed in a future that ever receded? "So we beat on, boats against the current, borne back ceaselessly into the past."

This is how culture works, making us captives of our history. If we let America's Prairies and Crow Creeks go under, leaving most of us in decaying cities or the probably failed experiment of suburbia, Fitzgerald's current becomes shallower, slower, it loses its force. If our cities are to stay healthy, the Prairies and Crow Creeks must survive and thrive.

Not everybody shares my central concern with culture. Lester R. Brown of Washington's Worldwatch Institute, a colleague of over 20 years, and I used to argue about it. Years ago he wrote me, "I realize you emphasize the need to take cultural factors into account. Nonetheless, it is my conclusion that the era of cheap food is past and the real cost of expanding food production will rise, making it extremely difficult to eliminate hunger and malnutrition."

Brown, probably the most effective publicist of environmental issues— he would put us all back on bicycles in a solar-energized, reforested, recycled future and I'm sure he's right—has been reluctant until now to say that cultural adaptation is key. But in his *State of the World 1990* he says at last: "Movement toward a lasting society cannot occur without a transformation of individual priorities and values. Throughout the ages, philosophers and religious leaders have denounced materialism as a viable path to human fulfillment. . . . As public understanding of the need to adopt simpler and less consumptive lifestyles spreads, it will become unfashionable to own fancy new cars and clothes. This shift, however, will be the hardest to make. . . ."

Brown uses the word *lifestyles*. Earlier I defined *lifestyles* as a fairly new word coined to fit rich, urban, high-technology society's new freedom of choice in how it will live. Explicit, in Brown's words, is that lifestyles, like fashions, can come and go. But, as I have emphasized, culture and lifestyles are not the same thing. Lifestyles are something we can choose. Culture is largely inherited, evolves very slowly, and mainly changes in response to changes in its economic basis—in Prairie and for the farmers of Crow Creek this means tillage of land. Freedom of choice can mean throwing away a lot of the old restraints, religious conventions, and patterns of obedience to authority that underpin culture. But the values Brown, who grew up on a farm, would like to see return are essentially cultural ones rooted in our rural past.

We have heard ideas on what to do about the farm problem. As I was writing these final pages, I also asked, as I said, William H. McNeill, whose *Rise of the West* traces the role of agriculture down through the ages, for a long-term view of the cultural issue. Professor McNeill studied under anthropologist Robert Redfield and later became Arnold Toynbee's foremost disciple. In my village work I frequently draw upon the ideas of all three of them. He at once agreed that the decay of rural America is "indeed a critical factor in our society which attracts little attention." He said his newest book, *Population and Politics since 1750,* looks at other urban societies slowly being cut off from their rural roots. In the United States this has come rather quickly; as recently as 1920 nearly half of young Americans, he says, "grew up with the discipline and experience of farm life," compared with today's less than 2 percent.

Drawing on a lifetime of study, Professor McNeill puts this into perspective in his book, essentially a series of lectures he gave at the University of Virginia in 1988–89. He sees all civilizations as biologically and culturally growing or decaying, not standing still. He says the world's demography has been greatly transformed twice, once when villages first began to cultivate the soil, allowing for more people, and in modern times when medical science freed us from much disease.

"Rural life," Professor McNeill goes on, "may never have been as stable and unchanging as city dwellers like to imagine. But as long as village births and deaths almost matched one another, one generation could follow another on the land, forming new households in accord with age-old practices so as to maintain the village community as before—more or less. But when systematic population growth set in, rural society became a ticking time bomb. Sooner or later, the rising generation was sure to run out of local access to suitable land on which to pursue the even tenure of its ways. When that happened, behavior had to change, not just for an elite but for the majority—indeed for everybody. Human society is still staggering under this momentous departure from rural routines; and because the modern growth of population has by no means run its course, we must expect further upheavals, especially in view of the fact that some populations have stopped growing and will, if present trends continue, begin to shrink while others are still in full spate of growth. Such a juxtaposition of growing and shrinking populations creates a new context for human affairs."

Civilizations, he argues, have always depended on internal circulation of peasants into cities to supply urban society with menial workers, soldiers, and its cultural underpinning (stronger institutions of family and property) and biological underpinning (more babies). "Urbanization," he says, "remains inimical to child bearing and rearing in a way that rural living is not, and with modern methods of birth control young women are able to regulate births to suit themselves."

He wonders what will happen without "the age-old pattern of the demographic circulation between town and country, rich and poor, upon which civilized society has depended since the third millenium B.C." Already in European nations where modern farm technology has led to a sharp drop in rural populations, city people are failing to replace themselves. (When enough married couples decide to have no more than two children, the overall population starts to shrink because many people have no children at all.) These diminishing numbers among wealthier, urbanized people have been offset the past 20 years only by massive and growing migration by peasants across political and cultural borders, such as Hispanics into the United States, Turks into Germany, Algerians and Moroccans into France, West Indians and Pakistanis into Britain. (McNeill wonders if the meek won't someday inherit the earth.)

He says we are on the cutting edge of time, reminding me of Barbara Ward's earlier quoted phrase of 20 years ago, "bordering on the edge of time." "The human majority," Professor McNeill writes in *Population and Politics since 1750*, "is stirring around the entire globe for the first time in history. This makes our age different from any that has gone before." He disturbingly reminds us, "Reproduction, both biological and cultural, is not automatic and unchanging."

When I asked McNeill if America's falling rural population meant cultural decline, not just for small towns and farms, but for our cities too, he agreed it did. McNeill: "The very real virtues and discipline of work that farming required is totally missing from all but a tiny fraction of our people. What this means, I think, is that the age-old way by which tradition and habit was transmitted from generation to generation has been interrupted; and in urban contexts, where the family is not the unit of work or of production, the nurture of the young and the transmission of culture has in practice been shifted from the bosom of the family to a far looser and less efficient network of human relations.

"Indeed I suspect it is mainly a youth cohort that educates its members to its own norms: and some of these norms are if not anti-social at least incompatible with long-term economic prosperity and hard work. Instant gratification and a refusal to subordinate one's personal impulse to any larger social solidarity is the way much yuppie behavior strikes me; and I do not think a society can thrive or long endure on such a basis."

All of this is connected with the monetization of work, he says, affecting women as well as men in our cities. "A devaluation of nurturing results. Farming had the virtue of keeping man and wife on the scene within the family context where the young had male and female models of adult behavior to internalize; and where youthful rebellion was difficult, since there was minimal support—group support—for rebellion."

In urban America today, McNeill continues, this is carried to the other extreme. "In town the youth gangs of our streets provide all the support anyone needs for rebellious behavior; and in some contexts it becomes so entrenched as to defeat formal education and produce young men who are unfitted for anything but criminal activity. This is the most critical issue for our society, I believe, and one we have no way of coping with yet." Professor McNeill feels some sort of youth corps, something like the old CCC of the New Deal's Depression days, militarized and disciplined in army fashion, might be needed. "That can work: but it is a far cry from the looser, more individualized, and 'voluntary' sort of apprenticeship to adult life that farm families provided in times past."

Isn't it true, I asked, that Third World cities are as stable as they are because so many of their poorest people are uprooted villagers who still cling to their rural cultures? He agreed, but added, "The parasitic city is still with us also. Recruitment is simply coming from afar and across cultural, linguistic, and racial lines." For example, he said, many American service jobs are now held by people who grew up in villages but speak

Spanish or another foreign tongue. "Cities have always been places where biological wastage took place: once mainly due to diseases and infections, now due to birth control and drugs and other forms of self-destruction. Work habits that are sequestered from the place children exist cannot be transmitted to the young very effectually. Habits that go along with the work and are necessary for its effectual accomplishment are difficult to preserve when work leaves the family context. That perhaps is the problem of city living—at least in our time."

If you visit old marketing and handicraft centers in Third World cities like Cairo or Delhi, with their narrow streets and ancient houses and charm, if bad sanitation, you can see what he means. You find large numbers of traditional artisan families who still work together, making and selling wares at home; invariably local religious and moral norms are strong.

"Once when artisan work prevailed," Professor McNeill concludes, "family structures could flourish there too; and it was the intensification of infection that made the cities sinkholes for human life. Now it is the incompatibility of city living with effective transmission of culture from adult to young that seems to me to stand in the way of a stable urban culture.

"This is, of course, worldwide: other urban cultures also are seeing biological and cultural extinction coming close. No substitute for the rural base of urban society has yet been invented."

PROFESSOR MCNEILL'S PROFESSIONAL field of vision is world history, just as Dr. Borlaug's is world agriculture, Lester Brown's the world's environment, and my own the world's villagers and what, in an anthropological sort of way, they have to tell us. But history makes no sense without agriculture, nor agriculture without the environment, nor all three without ordinary people going about their lives. Which brings us back to Prairie and Crow Creek.

Our horizons tend to extend only as far as we ourselves are likely to travel. To a Third World villager this can even mean just a few miles in any direction. War broadened Bud Petrescu's horizons to Vietnam, Adam Pierce's to Korea. Lionel Farrow's go across the Rio Grande, Mrs. Muhlenthaler's as far as a leper colony in the Cameroon jungle.

But for most people, in Prairie and Crow Creek and elsewhere, there is not too much curiosity about or knowledge of the greater world. Television is changing this, but a hometown still remains the fixed point by which a man or woman knows his or her position in the world and relationship with the rest of humanity.

You and I might not want to spend the rest of our lives in Prairie or Crow Creek, but a remarkable number of their people do. For good reason. I think they sense this is the most natural way for human beings to live. They know there is a negative side to it; that if they stay at home, it means accepting the restraints and conventions necessary to people who live in groups. Fear of neighbors' censure or "what will people say" is a potent force; we've seen the pressure to conform to old ways, go to the Civic League, forsake the pot-smoking beer party. This is reinforced by gossip—still, despite cable TV and videos, the chief form of entertainment. Our interviews would have been far less informative if rural people were not so intensely interested in each other's character and affairs.

Every rural community has its own system of mutual rights and obligations. Far from the monetized work of the cities that Professor McNeill talks about, this is giving labor for a common good without pay. It is understood that everybody has a part to play in an organic whole. Pedro Alvarez, who is closer, with his Hispanic background and Catholicism, to the traditional society, responded on the matter of the streetlights in terms of rural culture. Those who stayed aloof or refused to pay were all commuters.

These urban invaders—and many ex-farming communities will have to depend on them to survive—while they don't understand the rules, do want some of the old traditions for their children. They take part in Crow Creek's yearly Easter egg hunt, its community Christmas tree, tricks or treats at Halloween. But the social life that counts in a rural community turns around births, marriages, and deaths, the church and the school. It takes engagement and time. It takes restraint of self-interest in favor of community. It takes the old economic basis of farming to make it happen.

# For Further Reading

Unless otherwise indicated, I have taken quotations from interviews and correspondence, most of it in 1987 and 1990, the main two periods of research. This list is an informal one of additional sources for readers who might want to pursue the subjects at greater length for themselves.

CALVIN L. BEALE, "Significant Recent Trends in the Demography of Farm People," Philadelphia Society for Promoting Agriculture, 5 May 1988; *A Taste of the Country; A Collection of Calvin Beale's Writings* (University Park: Penn State Press, 1989); "Nonmetro Population Trends: Optimistic Note Ends Pessimistic Decade," USDA paper (figures up to mid–1988); "Agricultural Communities: Economic and Social Setting," testimony to Committee on Agriculture, U.S. House of Representatives, 1983; "Metropolitan and Nonmetropolitan Growth Differentials in the United States since 1980," USDA paper (despite title a review of 1970s going up to 1983); "Selected Aspects of Agricultural Communities and People in the United States," Congressional Research Service, 19 May 1983.

NORMAN E. BORLAUG, "World Revolution in Agriculture," *1988 Britannica Book of the Year* (Chicago: Encyclopaedia Britannica, 1988), coauthored with Christopher R. Dowswell; "Challenges for Global Food and Fiber Production," Royal Swedish Academy of Agriculture and Forestry, 1988; Convocation Address, Punjab Agricultural University, Ludhiana, India, 1987. Transcripts of my own interviews with Dr. Borlaug in Mexico, Texas, and Iowa since 1977 run several hundred pages.

LESTER R. BROWN, *State of the World* series (New York: W. W. Norton & Company, annually 1984–90), coauthored with his Worldwatch Institute staff but Brown writes the lead articles.

GILBERT C. FITE, *American Farmers: The New Minority* (Bloomington: Indiana University Press, 1981), the best historical work I have seen on the rise and decline of a large American farming population.

NEIL E. HARL, *The Farm Debt Crisis of the 1980s* (Ames: Iowa State University Press, 1990); "Nine Principles and Three Questions for Thinking about Agricultural Overproduction," *Choices*, Second Quarter 1987. Dr. Harl has also written an interesting paper on Poland, where he spent some time in 1990, "Poland: A Country in Transition," March 1990.

WILLIAM H. MCNEILL, *The Rise of the West: A History of the Human Community* (Chicago: University of Chicago Press, 1963); *Arnold J. Toynbee, A Life* (Oxford: Oxford University Press, 1990); *Population and Politics since 1750* (Charlottesville: University Press of Virginia, 1990); *Mythhistory and Other Essays* (Chicago: University of Chicago Press, 1986); *The Pursuit of Power* (Chicago: University of Chicago Press, 1982); *Plagues and Peoples* (New York: Anchor Press/Doubleday, 1976). Dr. McNeill also wrote Toynbee's obituary for the Proceedings of the British Academy, London, vol. 63 (Oxford: Oxford University Press, 1977). His "The Care and Repair of Public Myth," together with my "Science and the Villager: The Last Sleeper Wakes," made up the two lead articles in the 60th anniversary issue of *Foreign Affairs* in the fall of 1982.

DON PAARLBERG, *Toward a Well-Fed World* (Ames: Iowa State University Press, 1988); *Farmers of Five Continents* (Lincoln: University of Nebraska Press, 1984); "Farm Ownership: The Deep Issues," paper, 24 October 1986.

WAYNE D. RASMUSSEN, "The Mechanization of Agriculture," *Scientific American,* Summer/Fall 1982. In 1987 I met Rasmussen, after earlier interviewing him in Washington, at a conference he had arranged on behalf of USDA on the Middle Western farming crisis in Fargo, North Dakota.

VERNON RUTTAN, "Biological and Technical Constraints on Crop and Animal Productivity," University of Minnesota, revised December 1989; "Views and Reviews 1983–1988," May 1989. Professor Ruttan's distinction between biological and mechanical technology first came to my attention when we both did articles for a special issue on the world food problem in *Transaction/Society* in September/October 1980, which Ruttan edited.

THEODORE W. SCHULTZ, "The Long View in Economic Policy: The Case of Agriculture and Food," International Center for Economic Growth, San Francisco, 1987; "Dealing with Economic Imbalances Between Industry and Agriculture," IEA World Conference, December 1986. An extremely prolific writer over the past 60 years. I have found his 1964 *Transforming Traditional Agriculture* and his 1975 *Economics of the Family: Marriage, Children and Human Capital* most useful.

I have listed the above authorities separately, as their ideas have influenced my work for many years. For instance, *The Rise of the West* helped inspire my village studies and initially helped determine their geographical location. Dr. Borlaug and Dr. Schultz gave generous help with my books *Villages* and *Those Days*. More recent interviews with Borlaug, Calvin Beale, and William McNeill figured in my last book, *An American Looks at Britain*. In their respective fields, these are among the wisest people I know.

Works also drawn upon for this book, or related to issues it raises:

JOSEPH AMATO, *When Father and Son Conspire: A Minnesota Farm Murder* (Ames: Iowa State University Press, 1988). My view that this crime grew out of abnormal psychology and not the rural crisis of the 1980s is quoted by Professor Amato on the cover.

ARISTOTLE, his idea that culture is decided by how we get our food is taken from *Politics, i,* 8. Many of his views in *Politics* and *Ethics* have contemporary relevance though he lived 384–322 B.C., showing how little human behavior has changed.

SANDRA S. BATIE and ROBERT G. HEALY, "The Future of American Agriculture," *Scientific American,* February 1983.

ELISABETH BUMILLER, *May You Be the Mother of a Hundred Sons: A Journey among the Women of India* (New York: Random House, 1990), excellent for the way this former writer of *The Washington Post*'s Style section relates Third World village culture to our own.

RICHARD CRITCHFIELD, the last three of my seven books have the most relevance: *Villages* (New York: Doubleday/Anchor Books, 1981, updated 1983, a new edition to be published 1991) probes for the universal qualities of Third World village life; *Those Days: An American Album* (New York: Doubleday/Anchor 1986, Dell 1987) focuses on agricultural and cultural change 1880–1940; and *An American Looks at Britain* (New York: Doubleday, 1990), looks at an urban culture largely cut off from its rural roots.

WILL DURANT and ARIEL DURANT, *The Lessons of History* (New York: Simon & Schuster, 1978), the last thin volume by perhaps the greatest popularizer of philosophic ideas in American publishing history, written with the help of his wife, gives great importance to agriculture.

GEORGE M. FOSTER, *Peasant Society: A Reader* (Boston: Little, Brown, 1967), coedited with Jack M. Potter and May N. Diaz; gives Foster's concept of "limited good" in small communities. In his foreword to my book *Shahhat, an Egyptian* (Syracuse: Syracuse University Press, 1978, Avon Books, 1980), Foster notes I emphasize my credentials as a journalist, and says, "In our quest to know how others live, many paths lead to understanding. For anthropologists, comprehension of peasant society has been deepened by the accounts of novelists, poets, historians, political scientists, sociologists, psychologists—and now by that of a journalist."

IAN FRAZIER, "Great Plains," three-part series as "A Reporter at Large" in *The New Yorker,* February 20, February 27, and March 6, 1989. *Great Plains* subsequently published in book form.

THOMAS JEFFERSON, *Notes on the State of Virginia* (1787), his views on the importance of farming and rural culture on American character and society.

HOWARD KOHN, *The Last Farmer* (New York: Summit Books, 1988, reprinted by Thorndike Press, 1989), a moving portrayal of a son with an urban culture coming to appreciate a father with a rural culture.

MARK KRAMER, *Three Farms: Making Milk, Meat and Money from the American Soil* (Cambridge: Harvard University Press, 1980, updated and reissued, 1987), another look at contemporary farming.

RICHARD LEAKEY, *The Making of Mankind* (New York: E. P. Dutton, 1981), his views on how human behavior is culturally determined.

OSCAR LEWIS, *Five Families* (New York: Basic Books, 1959), provides practical definitions of culture and shows how it works.

RICHARD LINGEMAN, *Small Town America; A Narrative History 1620–the Present* (New York: G. P. Putnam's Sons, 1980), magisterial, the definitive work.

ROBERT S. LYND and HELEN MERRELL LYND, *Middletown: A Study in Modern American Culture* (New York: Harcourt Brace Jovanovich, 1929), groundbreaking when published, largely of historical interest today.

NORMAN MACRAE, "America's Third Century," a survey, *The Economist*, 25 October 1975; as in his 1968 survey "The Neurotic Trillionaire," no one has written about America better.

ANDREW H. MALCOLM, *Final Harvest: An American Tragedy* (New York: Times Books, 1986), takes the opposite view about the same crime as the Amato book. His publisher claimed the 1983 killing of two Minnesota bankers by a bankrupt farmer and his son was "dramatic evidence of the death of not only individuals but, in large part, of the small-town farmer." In the book Malcolm, a highly respected reporter for *The New York Times*, did not go this far. In a dispatch he reported "a fundamental restructuring across the country's midsection, which historically has produced so much of the nation's food and factories, its leaders and social values." In my review for *The Washington Post*, I wrote: "Economists say it is getting too expensive to subsidize the American family farmer just to preserve our rural values. The truth, as *Final Harvest* illustrates, is that rural values—like respect for property or the place given marriage and the family—just may be our only values. As history shows, every society goes into decline, however slowly, once it gets too far away from its agricultural origins." It was one of the first times I raised the theme that developed into this book.

MARGARET MEAD, *Village Viability in Contemporary Society*, ed. Priscilla Copeland Reining and Barbara Lenkerd (Boulder: Westview Press, 1990); in a 1978 essay shortly before her death, Mead raised the possibility of an "elective village," combining the seeming contradiction of a rural cultural base with late 20th century freedom of choice.

DANIEL PATRICK MOYNIHAN, *Family and Nation* (New York: Harcourt Brace Jovanovich, 1986), a politician and intellectual gadfly who takes a rather anthropological point of view.

PRAIRIE FIRE RURAL ACTION, INC., "The Continuing Crisis in Rural America; Fact vs. Fiction," May 1987, an example of rural activism at the height of the 1980s crisis.

ROY L. PROSTERMAN and JEFFREY M. RIEDINGER, *Land Reform and Democratic Development* (Baltimore: Johns Hopkins University Press, 1987); as the dust jacket quotes me: "This is a definitive study of land reform in the late twentieth century and, I believe, a book of supreme importance."

ROBERT REDFIELD, *The Little Community* (1953) and *Peasant Society and Culture* (1956) (Chicago: University of Chicago Press, 1956), originally published separately, the two best works, I believe, ever done in American social anthropology and indispensable when it comes to understanding the nature of urban and rural culture and their relationship; also "The Cultural Role of Cities," *Economic Development and Social Change,* vol. 3, 1954, coauthored with Milton B. Singer.

RICHARD RHODES, *Farm: A Year in the Life of an American Farmer* (New York: Simon & Schuster, 1989), good in its mastery of the technical side of farming.

E. C. A. RUNGE, "Agricultural Productivity as the Solution Rather than the Problem," produced by a group of Texas A&M professors to brief Bush government leaders and members of Congress, 1989, a briefing document on expanding ethanol production from corn.

MARGARETHE ERDAHL SHANK, *The Coffee Train* (New York: Doubleday, 1953), fictional portrait of the town I've called Prairie as it was in the 1920s.

ALICE F. SKELSEY, *Biotechnology in Agriculture,* Joint Council on Food and Agriculture Sciences, U.S. Department of Agriculture, 1984, updated in newspaper clipping file, mainly *The Economist* and *New York Times* articles.

KENNETH E. STONE, "Impact of the Farm Financial Crisis on the Retail and Service Sectors of Rural Communities," paper, Department of Economics, Iowa State University, 1987, since updated. In October 1989, Professor Stone did a study of the impact of Wal-Mart stores on rural Iowa.

TONY THOMAS, "World Champions: A Survey of American Farming," *The Economist,* 5 January 1980.

TIME, "The New U.S. Farmer," cover story, 6 November 1972; "Going Broke: Tangled Policies—Failing Farms," cover story, 18 February 1985. *Time* magazine's interpretation of the farming boom of the 1970s and hard times of the 1980s.

ARNOLD TOYNBEE, *Civilization on Trial* (Oxford: Oxford University Press, 1946); *A Study of History* (Oxford: Oxford University Press, 12 volumes, 1934–61). A 617-page condensation of Toynbee's great work came out in 1947 and became a best-seller in America. "What Toynbee did in my view,"

William McNeill told me in 1988, "was to enlarge the field of history to make it embrace the whole of humanity."

LARRY WOIWODE, *Beyond the Bedroom Wall* (New York: Farrar Straus Giroux, 1975), 14 chapters previously appeared in *The New Yorker* beginning in 1965, another novel set in what has been called Prairie's Hawk County.

# Index

257

# About the Author

RICHARD CRITCHFIELD is the author of *An American Looks at Britain; Those Days; The Long Charade;* and a village trilogy *Villages, Shahhat,* and *The Golden Bowl Be Broken.* In 1981 he was awarded a MacArthur Fellowship. He also has received awards from the Overseas Press Club of America for his war reporting in Vietnam, and the Rockefeller Foundation, the Ford Foundation, and the Alicia Patterson Foundation. He is presently doing research in Poland, Russia, China, and India for a new village book.

# Also Available
# From Island Press

*The Living Ocean: Understanding and Protecting Marine Biodiversity*
By Boyce Thorne-Miller and John G. Catena

*Natural Resources for the 21st Century*
Edited by R. Neil Sampson and Dwight Hair

*The New York Environment Book*
By Eric A. Goldstein and Mark A. Izeman

*Overtapped Oasis: Reform or Revolution for Western Water*
By Marc Reisner and Sarah Bates

*Permaculture: A Practical Guide for a Sustainable Future*
By Bill Mollison

*Plastics: America's Packaging Dilemma*
By Nancy A. Wolf and Ellen D. Feldman

*The Poisoned Well: New Strategies for Groundwater Protection*
Edited by Eric Jorgensen

*Race to Save the Tropics: Ecology and Economics for a Sustainable Future*
Edited by Robert Goodland

*Recycling and Incineration: Evaluating the Choices*
By Richard A. Denison and John Ruston

*Reforming The Forest Service*
By Randal O'Toole

*The Rising Tide: Global Warming and World Sea Levels*
By Lynne T. Edgerton

*Rush to Burn: Solving America's Garbage Crisis?*
From *Newsday*

*Saving the Tropical Forests*
By Judith Gradwohl and Russell Greenberg

*War on Waste: Can America Win Its Battle With Garbage?*
By Louis Blumberg and Robert Gottlieb

*Western Water Made Simple*
From *High Country News*

*Wetland Creation and Restoration: The Status of the Science*
Edited by Mary E. Kentula and Jon A. Kusler

*Wildlife and Habitats in Managed Landscapes*
Edited by Jon E. Rodiek and Eric G. Bolen

For a complete catalog of Island Press publications, please write:
  Island Press
  Box 7
  Covelo, CA 95428
  or call: 1-800-828-1302